Seismic Response of Buried Pipes and Structural Components

A report prepared by the Committee on Seismic Analysis of the ASCE Structural Division Committee on Nuclear Structures and Materials

AMERICAN SOCIETY OF CIVIL ENGINEERS FOUNDED 1852 ®

Published by the
American Society of Civil Engineers
345 East 47th Street
New York, New York 10017-2398

The material presented in this publication has been prepared in accordance with generally recognized engineering principles and practices, and is for general information only. This information should not be used without first securing competent advice with respect to its suitability for any general or specific application.

The contents of this publication are not intended to be and should not be construed to be a standard of the American Society of Civil Engineers (ASCE) and are not intended for use as a reference in purchase specifications, contracts, regulations, statutes, or any other legal document.

No reference made in this publication to any specific method, product, process, or service constitutes or implies an endorsement, recommendation, or warranty thereof by ASCE.

ASCE makes no representation or warranty of any kind, whether express or implied, concerning the accuracy, completeness, suitability or utility of any information, apparatus, product, or process discussed in this publication, and assumes no liability therefor.

Anyone utilizing this information assumes all liability arising from such use, including but not limited to infringement of any patent or patents.

Library of Congress Catalog Card No.: 83-71304
ISBN 0-87262-368-8
Manufactured in the United States of America.

SEISMIC RESPONSE OF BURIED PIPES AND STRUCTURAL COMPONENTS
REPORT BY THE SEISMIC ANALYSIS COMMITTEE OF THE ASCE NUCLEAR STRUCTURES AND MATERIALS COMMITTEE

Working Group for Seismic Response of Buried Pipes and Structural Components

P. K. Agrawal, Sargent & Lundy Engineers
E. C. Goodling, Gilbert Associates, Incorporated
R. C. Guenzler, EG&G Idaho, Incorporated
R. J. Hunt, Tennessee Valley Authority
I. M. Idriss, Woodward-Clyde Consultants
H. J. Kuo, Ebasco West
M. J. O'Rourke, Rensselaer Polytechnic Institute

ABSTRACT

The Seismic Analysis Committee* of the ASCE Nuclear Structures and Material Committee has prepared this technical position paper for the seismic response of buried pipes and structural components at nuclear power plants. The committee has documented the present practice of analysis and design and identified areas where additional investigations might be desirable. Additional work is needed to better define seismic waves, to document the behavior of long buried structures during earthquakes, and to develop design methodologies. Anyone having information, comments, or suggestions is requested to provide them to any of the committee members.

INTRODUCTION

At nuclear power plants there are numerous underground safety-related pipes and structural components which are utilized in the operation of the plant. These underground components provide water supply, electrical power, routing of pipes, and movement of personnel to ensure safe operation and safe shutdown of the plant. Examples of these are essential raw cooling water pipes, electrical conduit banks, pipe tunnels and trenches, and personnel tunnels.

*See Appendix 5 for list of the members of the Seismic Analysis Committee.

Figure 1 illustrates some typical cross sections of these buried components. These are constructed of steel or reinforced concrete, or combinations thereof, and are usually long structures supported continuously by the surrounding soil.

The underground safety-related pipes and structural components require seismic analyses and design to ensure the safety of the nuclear power plants. Briefly this involves layout and qualification of the route, consideration of nonseismic loads which influence layout, determination of seismic loads to be considered, seismic analyses, and seismic design. Each of these factors is considered in detail in the following sections.

QUALIFICATION OF ROUTE AND LOADINGS

General Considerations and Loadings

In investigating and evaluating the route of a proposed pipe or other buried component, it is useful to keep in mind the potential effects that an earthquake can have on a pipe and on structures adjacent to the pipe. The first, and potentially severe, effect would be abrupt displacement in a zone of earthquake fault breakage. However, because nuclear power plants are sited to avoid zones of active faulting, this potential effect would not generally require evaluation for a pipe route at a nuclear plant.

The second effect would be ground failure due to the seismic shaking. Included in a broad definition of ground failure are liquefaction, landsliding, lateral spreading, and gross settlement. The effects of these phenomena on a pipe can be severe in some cases. Thus, the potential for ground failure along the route requires evaluation even though nuclear plant sites having soils particularly susceptible to failure would be avoided in general during the site selection process.

Transient, recoverable deformations of the ground during seismic shaking constitute the third effect on buried pipes. Transient loads and deformations are induced in buried components by two phases of shaking. The first phase is related to seismic waves in the surrounding soil and the second phase is related to deformations of the buildings at connections of the buried components.

The first phase is the transient deformation imposed by seismic waves traveling through the soil surrounding the pipe or conduit. Such deformations include longitudinal tension, compression, and bending in a pipe resulting when ground motion at two points along the propagation path of the seismic waves are out-of-phase. In the case of large-diameter pipes, tunnels, and box structures, deformations of and/or lateral earth pressures acting on the cross section of the structural element may also require consideration in the design. In general, rigorous time-dependent analytical treatment of this transient seismic motion is not considered practical. The overall complexity of

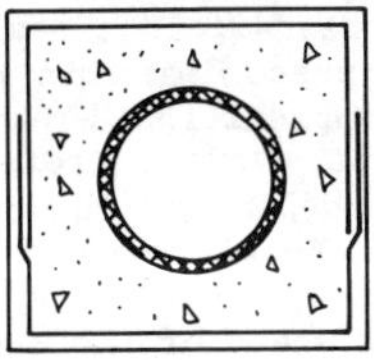

CONCRETE ENCASED
STEEL PIPE

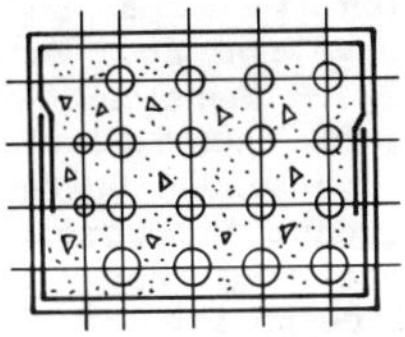

ELECTRICAL CONDUIT
BANK

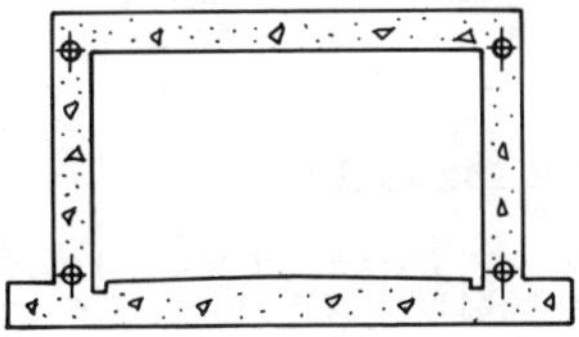

PIPE TUNNEL

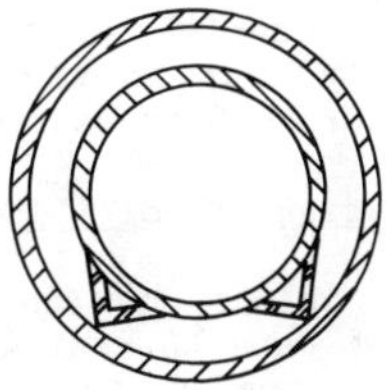

STEEL PIPE ENCASED
STEEL PIPE

FIGURE 1 - TYPICAL CROSS SECTIONS OF BURIED PIPES AND STRUCTURAL COMPONENTS

the earthquake mechanism makes it difficult to predict the seismic wave type that causes the peak ground motions and corresponding strain levels at a location given the assumptions of this paper and neglecting limitations of axial strain by friction. In general, shear (s) and compression (p) waves are predominant in near-field distances, whereas the Rayleigh and Love surface waves are predominant in the far field (figure 2). There are exceptions and complexities in the near field depending on the direction from the fault surface (reference 1) and depth of overburden (reference 4). Focal depths of the earthquake can also influence the waves. Increasing the focal depth tends to reduce the surface wave energy relative to the body wave energy. Increasing the focal depth also increase the apparent speed of the body waves with respect to the pipeline. Based on the complexities of the seismic wave types and combinations, conservative design procedures have been adopted to assure that the total stresses or strains remain within basic design allowables. These procedures are discussed later under Determination of Earthquake Loads.

The second phase of shaking from the transient motions is the effect of seismic-induced motions and deformations of structures or buildings to which the pipe is attached or the settlement of earthfill adjacent to the building. In locations where piping enters a structure, additional loads will be imposed by differential building movements. These movements are computed and imposed on the piping end points. Where the piping terminates at two different structures, building motion is conservatively assumed to be out of phase. Frequently, designs of terminal points include rattle space to provide flexibility that accommodates relative motion. Figures 3 and 4 are examples of such protective provisions.

Investigation of Soil Conditions Along the Route

The intent of the investigation phase should be to gather sufficient data to evaluate the acceptability of the route and provide parameters needed for the design of the pipe. The general soil conditions along the route can be established by conventional programs of soil borings and laboratory tests of soil samples. Of particular interest in the investigation are the presence and extent of soils that have a high potential for liquefaction or densification (particularly loose sands), changes in the soil conditions along the route that may increase potential for differential permanent or transient displacements of a pipe, and proximity of slopes that may create a potential for lateral ground movements.

Soil properties of general interest include grain size and plasticity characteristics, density and relative density, and static strengths. To establish resistance of the soils to liquefaction or deformation, supplementary cyclic loading tests should be considered. Field standard penetration tests (SPT) or cone penetration tests (CPT) also provide valuable information on the general soil characteristics and their resistance to liquefaction and deformation.

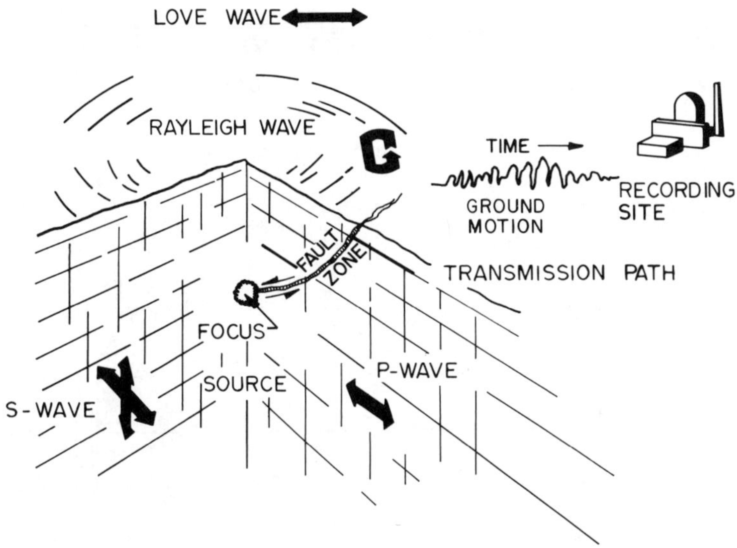

FIGURE 2- SCHEMATIC ILLUSTRATION OF SEISMIC WAVES AFFECTING UNDERGROUND PIPING SYSTEMS. P-WAVES ARRIVE AT SITE FIRST, FOLLOWED BY THE S-WAVES AND THEN THE SURFACE WAVES.

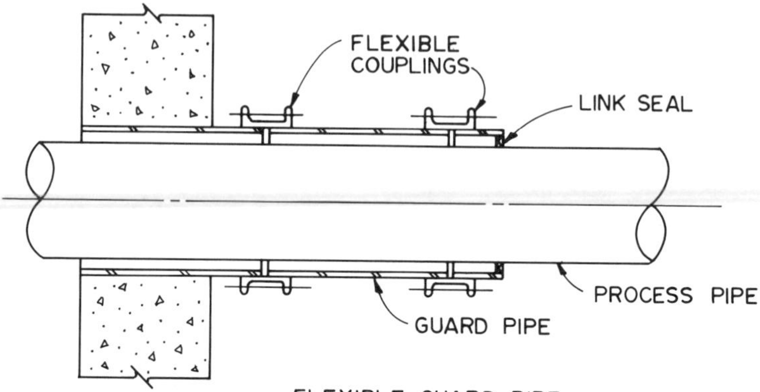

FLEXIBLE GUARD PIPE

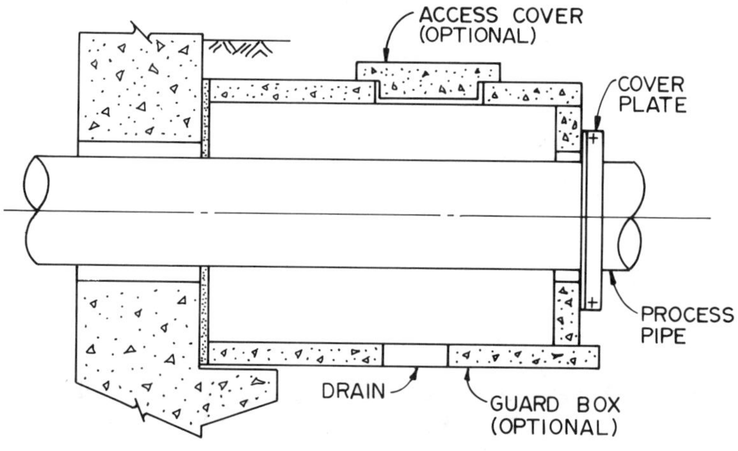

REINFORCED CONCRETE GUARD BOX

FIGURE 3-BUILDING INTERFACE DETAILS

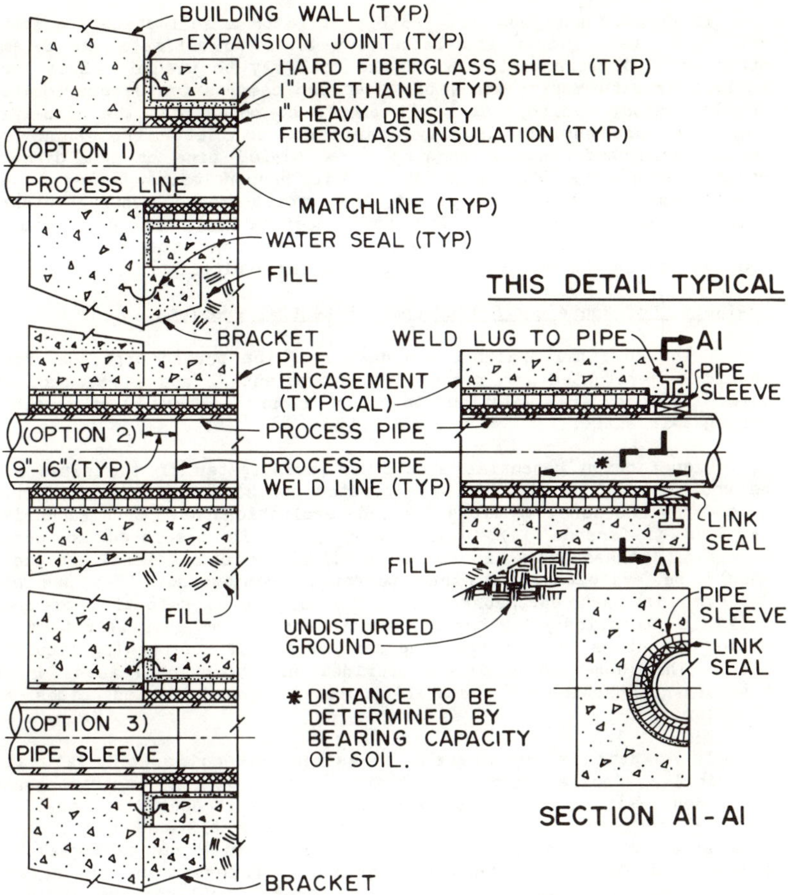

FIGURE 4 - OPTIONAL ENCASEMENT DETAILS FOR BURIED PIPE BUILDING PENETRATIONS

If dynamic analyses of a soil profile or a soil-structure system are to be made, field and laboratory tests to estimate the dynamic stress-strain characteristics of the soil may be needed. Such tests include in-situ measurements of shear and compression wave velocities and laboratory cyclic tests. It should be noted that the simplified analysis procedures usually employed to determine transient longitudinal and bending strains in a buried pipe do not directly require knowledge of the soil dynamic properties. Rather, these procedures utilize estimates of the ground motion characteristics (e.g., peak particle velocity and acceleration) and estimates of apparent traveling wave velocities resulting from complex wave propagation in the soils and rock underlying a site.

Evaluation of Route and Solutions to Potential Problems

Prior to final analysis and design of the buried pipe, the route should be evaluated for ground failure potential and effects on the pipe. If necessary, solutions to potential problems would be developed during this stage.

Liquefaction potential should be investigated if the pipes will be underlain or surrounded by saturated cohesionless soils. Current procedures and considerations for such evaluations have been summarized by Seed (reference 19). If liquefaction is predicted to occur, potential consequences including settlements and lateral movements should be evaluated. Another potential consequence that has been observed in past earthquakes due to complete liquefaction of loose soils is flotation of a buried structure when the unit weight of the buried structure is less than the unit weight of surrounding liquefied soil. This observed condition provides one important reason for the adequate compaction of backfills placed around buried pipes and structures.

In partially saturated cohesionless soils or saturated cohesionless soils that can drain freely during an earthquake, settlements may occur during the earthquake due to densification of the soil by the ground shaking. Post-earthquake settlements may occur if pore water pressures are generated in soils during the earthquake and dissipated following the earthquake. Information on seismic settlements of soils is provided in several publications including Seed and Silver (reference 18), Youd (reference 26), Lee and Albaisa (reference 7), and Pyke et al (reference 16). The magnitude of differential settlement of a buried pipe depends on the variation of subsurface conditions along the route and possible connections of the pipe to nonsettling structures.

The potential for lateral spreading or landsliding requires evaluation where buried pipes and structural components are located near slopes. The potential for lateral movements is substantially increased if the soils soften significantly due to seismically induced pore pressures in the soils.

If unacceptable pipe behavior is predicted to occur along the route due to ground failure phenomena, various solutions to the potential problems may be considered. Some of these solutions include: (a) rerouting to avoid areas of problem soils, (b) removal and replacement of soils, (c) soil stabilization (e.g., by densifying, grouting, and draining), (d) supporting the pipe or structure in soils not susceptible to failure (e.g., by deeper burial or pile foundations extending into stable soils), and (e) designing to accommodate limited ground movements, either by providing flexibility to adjust to the ground displacements or by providing rigidity to resist the ground displacements. However, the latter may be difficult.

Other Considerations Governing Route Selection

The route selection for buried structural components should include protection from other natural phenomena and localized loads. Route selection should avoid areas of, or provide protection from turbine missiles, moving loads (roadways, cranes, etc.), and excessive overburden loading. Route selection should also consider the proximity and effects of other large-scale structures that can affect soil motion and soil pressures surrounding the pipe. The effects of the collapse of non-Category I structures should be considered in selecting the route. Tornado missiles that have a definite effect on the depth of burial should also be considered.

DETERMINATION OF EARTHQUAKE LOADS

As discussed above, the seismic waves causing the loads on a long buried structure are extremely complex. To completely describe the behavior of the structure would require a 3-dimensional dynamic analysis of the structure and surrounding soil, including the effects of the soil-structure interaction, earthquake source characteristics, and transmission paths of the different seismic waves. Since this type of rigorous analysis is seldom economically justified, a simplified approach is considered reasonably conservative and acceptable.

If the following assumptions are made about the nature of the seismic event and its interaction with a long buried structure, expressions derived by Newmark (reference 11) and expanded on by Yeh (reference 25) determine the instantaneous axial strain and bending strains at a section of the buried structure due to compressional, shear, and surface waves.

Assumptions:

1. Each wave (compressional, shear, or surface) is propagated in one direction without interference from the other waves.

2. The changes in the wave shape from one point to another as the wave advances are ignored.

3. The relative motion of the buried structure is the same as the surrounding soil except at the elbows or tees or if the friction force between pipe and soil is exceeded; the buried structure is assumed flexible.

<u>Correlation of Observed Earthquake Behavior with Assumptions of Simplified Analysis</u>

Some documented cases of observed earthquake behavior and tests results support the assumptions made for the simplified analysis. They are:

a. Tamura (reference 22):

From model tests one can look at tunnels as a beam without mass connected to ground.

b. Kubo (reference 6):

A number of observations indicate that the response of a buried pipe is strongly governed by deformation of surrounding ground.

c. Nakayama et al (reference 9):

Field observations of a 2-lane highway tunnel.

For two earthquakes with large epicentral distances, long period motions predominated and the tunnel had the same behavior as ground. For two earthquakes with small epicentral distances, short period components predominated and the tunnel behavior differed slightly from that of ground.

d. Sakurai and Takahashi (reference 17):

Observations for 13 earthquakes with magnitude 5; the ground strain was measured by earthstrain meters.

No observed difference between the pipeline and ground deformation.

Axial strain predominated.

Maximum strain in pipeline did not occur at the same time of maximum ground acceleration but occurred afterwards.

e. Nasu et al (reference 10):

Experimental tests were made using artificially generated ground motion. Pipe and ground were observed to move together as long as adhesion was not lost.

f. Okamoto et al (reference 13):

Observation of 13 meters by 7.5 meters railroad tunnel.

The strain in the tunnel was produced by deformation of ground and not by inertia forces.

Low frequency predominated the strain records.

g. Yeh (reference 25):

Investigated behavior of a storm drainage tunnel during the San Fernando 1971 earthquake to correlate cracking with analysis methods.

General Equations for Axial and Bending Strains

With the above assumptions, the instantaneous axial and bending strains in a section of pipe are given by

$$\varepsilon_a = \pm \frac{v}{c} \qquad (1)$$

$$\varepsilon_b = \pm \frac{Ra}{c^2}$$

where:

ε_a = Axial strain

ε_b = Bending strain

v = Particle velocity

a = Particle acceleration

c = Apparent wave velocity with respect to ground surface

R = Radius of pipe

For an incident wave with velocity c , the apparent wave velocity is given $\frac{c}{\cos\theta}$, where θ is the incident angle of the wave, figure 5.

Strains Due to Compression Waves

For an incident compressional wave with an apparent velocity c_p and displacement amplitude A_p, the axial and bending strains in the pipe are given by

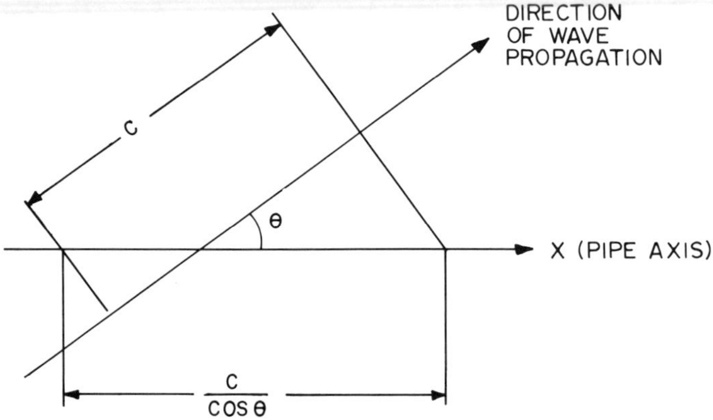

PLAN VIEW

FIGURE 5 - WAVE VELOCITY IN A PIPE DUE TO AN INCIDENT WAVE

$$\varepsilon_a = \pm \frac{v_p}{c_p} \cos^2 \theta$$

$$\varepsilon_b = \pm \frac{Ra_p}{c_p^2} \sin \theta \cos^2 \theta \qquad (2)$$

where v_p is the particle velocity and a_p the particle acceleration due to the compressional wave, figure 6a. The maximum axial and bending strains are given by

$$\varepsilon_a = \pm \frac{v_p}{c_p} \quad ; \theta = 0^\circ \qquad (3)$$

$$\varepsilon_b = \pm 0.385 \frac{Ra_p}{c_p^2} \quad ; \theta = 35^\circ\text{-}16'$$

Strains Due to Shear Waves

For an incident shear wave with an apparent velocity c_s and displacement amplitude A_s, the axial and bending strains in the pipe are given by

$$\varepsilon_a = \pm \frac{v_s}{c_s} \sin \theta \cos \theta$$

$$\varepsilon_b = \pm \frac{Ra_s}{c_s^2} \cos^3 \theta \qquad (4)$$

where v_s is the particle velocity and a_s the particle acceleration due to the shear wave, figure 6b. The maximum axial and bending strains are given by

$$\varepsilon_a = \pm \frac{v_s}{2c_s} \quad ; \theta = 45^\circ \qquad (5)$$

$$\varepsilon_b = \pm \frac{Ra_s}{c_s^2} \quad ; \theta = 0^\circ$$

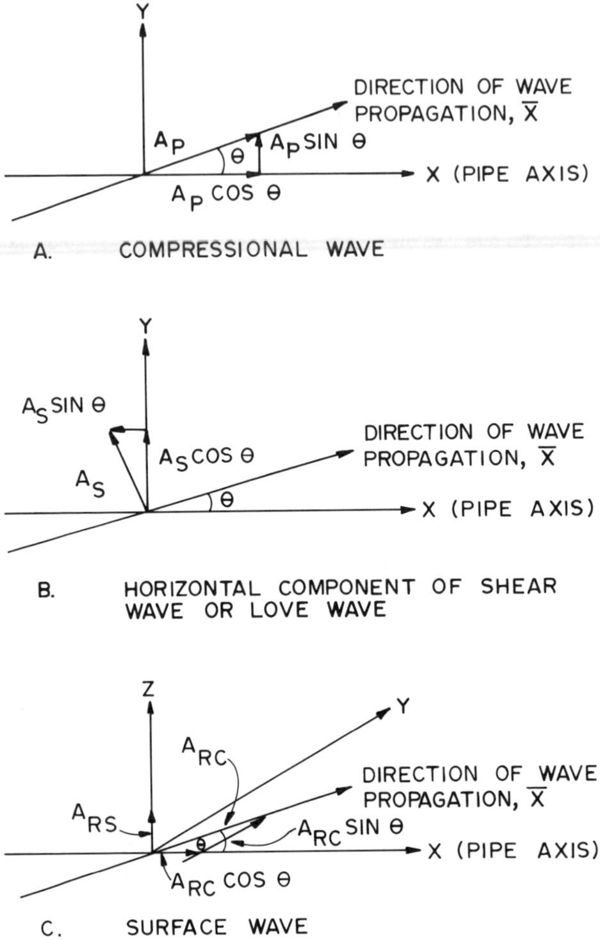

FIGURE 6 - AMPLITUDES OF COMPRESSIONAL, SHEAR AND SURFACE WAVES (REF. 9)

Strains Due to Surface Waves

For an incident surface wave with an apparent velocity c_R and displacement amplitude A_R equivalent to a compressional wave with amplitude A_{Rc} and a shear wave with amplitude A_{Rs}, figure 6c, the axial and bending strains in the pipe are given by

$$\varepsilon_a = \pm \frac{v_R}{c_R} \cos^2 \theta \qquad (6)$$

$$\varepsilon_b = \pm \frac{Ra_R}{c_R^2} \sin \theta \cos^2 \theta \quad ; \text{ compressional component}$$

$$\varepsilon_b = \pm \frac{Ra_R}{c_R^2} \cos^2 \theta \quad ; \text{ shear component}$$

where v_R is the particle velocity and a_R the particle acceleration due to the surface wave. The maximum axial and bending strains are given by

$$\varepsilon_a = \pm \frac{v_R}{c_R}$$

$$\varepsilon_b = \pm 0.385 \frac{Ra_R}{c_R^2} \quad ; \theta = 35^\circ\text{-}16' \quad ; \text{ compressional component} \qquad (7)$$

$$\varepsilon_b = \pm \frac{Ra_R}{c_R^2} \quad ; \theta = 0^\circ \quad ; \text{ shear component}$$

Consideration of Axial and Bending Strains

An examination of equations 3, 5, and 7 shows that the axial strains will be considerably larger than the bending strains for most long buried structures. Shearing strains also exist for long buried structures, but unless the soil is extremely stiff, the shearing strains probably can be ignored. As discussed above, these strains are based on assuming the motion of the buried structure is the same as the surrounding soil, that is, the relative motion between the pipe and soil is zero. In some cases this is not a good assumption, particularly for the axial strain. The axial strain has an upper limit depending primarily on the friction between the soil and the structure. Shah and Chu (reference 20) and Sakurai and Takahashi (reference 17) have outlined methods for considering the friction between the soil and the structure. Sakurai and Takahashi (reference 17) indicate that the upper bound axial force for a straight pipe is equal to the friction force per unit length times one-quarter of the wave length. The wave length can be evaluated by determining the natural period from a site-specific analysis.

Consideration of Strains Due to Different Wave Types

The above equations for strain are specifically for the different seismic wave velocities (c_p, c_s, c_r) and the associated ground velocities (v_p, v_s, v_r) and accelerations (a_p, a_s, a_r) for these waves. As discussed earlier, the overall complexity of the earthquake mechanism makes it difficult to predict the wave type that will cause the maximum ground velocity or acceleration. No specific measurements of the ground velocity or acceleration due specifically to the s, p, or r waves have been made. Only measurements of the maximum ground velocity and acceleration, regardless of the wave type or combination, are obtained. Also orthogonal components of each wave type could result in three components of like strain on a cross section of the pipe. Again, because of the complexity of the waves and the wave type, it is difficult to assign velocities or accelerations to the orthogonal components of each wave type. Therefore the maximum ground velocity and acceleration are normally used in equations 2 through 7. Combined with the use of the conservative wave propagation velocities discussed below, this is considered adequately conservative and eliminates the need to combine orthogonal components of each wave type by the square root of the sum of the squares or other methods.

Because of the different velocities and attenuation characteristics of the different wave types and the conservatism of the analysis approach it can be assumed that any section of a pipeline will experience the effects of only one type of wave at a given time. The choice of the wave type to be used for the design depends on the situation of the buried structure with respect to the earthquake source. For close distances to the fault the body waves, both shear and compression, would give larger amplitudes of motion, whereas at a large distance from the fault the body waves would have largely attenuated and the surface waves would govern the amplitudes of motion. Generally the compression wave does not control and can be ignored. Therefore, the shear or surface waves should be used, depending on the distance from the fault. There is no hard rule for what the distance should be, but it is generally thought that the shear waves would control if the site is within one fault (rupture) dimension.

Application of the strain equations 1 through 7 requires knowledge of the propagation speed of the seismic waves with respect to the pipeline. As discussed in the section on Qualification of Route and Loadings, the wave velocities of the overburden material can be determined from in-situ and laboratory tests. It is conservative to use the wave velocities of the soils surrounding the buried structure. However, as indicated by O'Rourke et al (reference 14) and Newmark and Hall (reference 12) it is most likely that this is overly conservative. Reference 14 shows that for a uniform soil layer overlying rock, the propagation velocity of seismic waves along a buried structure is found to be strongly influenced by the wave speed in rock. It also shows that the apparent speed of the shear waves in general is always greater than the speed of the waves in rock. Measured propagation velocities (references 23 and 24) also indicate that the speed of the seismic waves is much higher than the velocities of the soils immediately adjacent to the buried structures. The few values of measured apparent propagation velocities vary from about 8,000 fps (2440 m/sec) to 16,000 fps (4880 m/sec). These high propagation velocities are not appropriate to use in the simplified equations though, because at these high velocities the assumption of a constant wave shape may no longer be valid (O'Rourke, reference 14). Therefore, the design propagation velocity to be used in the simplified equations is somewhere between the velocities of the immediate surrounding soil and the apparent propagation velocities with respect to the ground surface.

A conservative and yet reasonable design value of apparent propagation velocity is 2,000 fps (610 m/sec) to 3,000 fps (915 m/sec) and the analyst should not be concerned with changes in the wave shape for component soils. These values should be used with equation 1 to calculate the strains on the buried structure for design, regardless of the type of wave that caused the velocities. For other values to be used, it is recommended that specialized assistance be sought in choosing the apparent design propagation velocity for each site.

Consideration for Bends and Localized Conditions

The assumptions used in deriving equation 1 are simplifications of the actual problem. The strain given by equation 1 is the localized strain in a section of a long buried structure and does not give the overall behavior of the structure. The effects of changes in the direction, size, or material of the structure are not considered. Also not considered is the behavior of the long buried structure where it is attached to a building. In their papers, Shah and Chu (reference 20) and Goodling (references 2 and 3) extended the application of equation 5 for incident shear waves to derive expressions for forces and moments at bends and connections to building; the expressions, however, are also applicable for incident compressional and surface waves. They considered the effects of the resistance of frictional forces and the passive resistance of the soil movements in long buried structures at bends and building connections.

At a bend, as shown in figure 7, the shears and moments (references 20 and 2) are given by

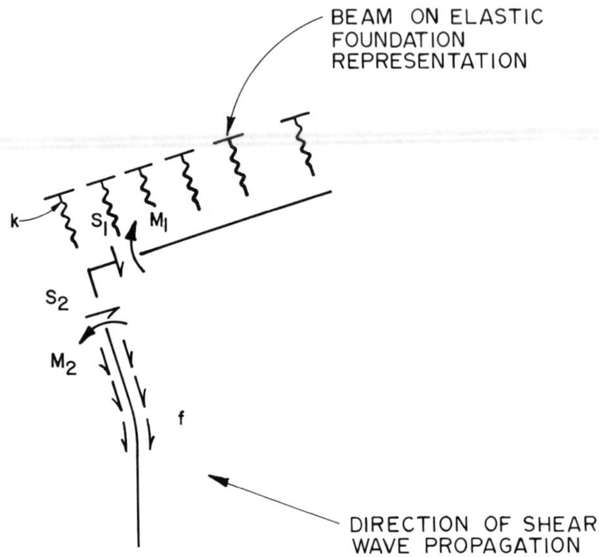

FIGURE 7 - SHEARS AND MOMENTS IN AN ELEMENT WITH A BEND

$$S_1 = \varepsilon AE - fL'$$

$$M_1 = M_2 = \frac{S_1}{3\lambda} \qquad (8)$$

$$S_2 = \frac{S_1}{3}$$

where:

S_1, S_2 = Shear in element

M_1, M_2 = Moment in element

ε = Axial strain as calculated by equations 1, 3, 5, or 7

A = Cross-sectional area of element

E = Modulus of elasticity for element

f = Frictional force per unit length of element

L' = Effective slippage length for element

$\lambda = \sqrt[4]{\frac{k}{4EI}}$; parameter for beam on elastic foundation

k = $k_o b$; k_o is the modulus of subgrade reaction for soil, b is the width of element on soil. Parmelee and Ludtke (reference 15) indicate that k is approximately equal to 3G, where G is the shear modulus of the soil.

Expressions for the effective slippage length, L', for long and short elements are given in reference 20. For elements rigidly connected to a building, the maximum axial force transmitted to the building will be

$$F_{max} = \varepsilon AE$$

Recommended values for the coefficient of friction to determine the frictional force are $\mu = 0.3$ to 0.5 for bare steel pipe embedded in normal soils. However, most pipes to be buried are coated with epoxy or coal tar. Based on test results given in reference 8, recommended values for coated pipes range from 0.5 to 0.9. The maximum value of friction is limited to the friction resistance of surrounding soil.

APPLICATION OF EARTHQUAKE LOADS IN DESIGN

Appendixes 3 and 4 contain a design example for a buried steel pipe and a concrete electrical duct bank, respectively. It is important to emphasize that seismic effects on buried structures are self-limited since deformations or strains are limited by seismic motions of the surrounding media. Following this philosophy, it is appropriate to treat them in much the same manner as the treatment of thermal effects on aboveground structures that are known to be self-limiting. The function of the underground structure should also be considered in the design. This is particularly true for underground concrete structures and determining reinforcement requirements. For example, can the concrete crack or how much cracking can be allowed and the structure still perform its function? Consideration of this question along with the functionality of the structure could change the design philosophy used.

There are two items of importance that should be pointed out concerning the design example for the buried steel pipe in Appendix 3. First calculations are made for the elbow considering it to be flexible and inflexible. The flexible condition should be the one used in design calculations. The inflexible condition is shown only for comparison purposes. Second, the ASME code check on the stresses is made considering both primary and secondary stresses. As discussed above, the seismic effects are self-limited and should be considered as secondary. The ASME code does not appropriately address buried piping, and its present use for buried piping as secondary stresses is unnecessarily conservative. The ASME code should be amended to properly address buried piping.

The general equations for axial and bending strain (equation 1) are based on the assumption that the seismic excitation can be modeled as a traveling wave whose amplitude and frequency content remain unchanged as it traverses the buried structure. Because of this, special design consideration should be made for sites with abrupt changes in the geotechnical supporting medium. Katayama (reference 5) noted unfavorable behavior of pipelines in soil consisting of mostly clay with spatial intrusions of hard soil. Using gelatin and silicone rubber models, Okamoto et al (reference 13) observed larger strains where the ground properties or the depth of the surface layer changes quickly. Finally an analytical study by Shinozuka and Kawakami (reference 21) has shown highest axial strains in regions where the soil conditions varied the most. Therefore, one should consider these situations in examining the route of the buried structure and make allowances for them in design if they cannot be avoided.

CONCLUSIONS

The present practice for analysis and design of buried pipes and structural components at nuclear power plants, using the simplifications and assumptions discussed above, is fairly straightforward. Still, several areas need more clarification. Some of these areas are (1) better definition of the seismic waves and their associated velocities for use in the analysis, (2) behavior of buried structure when it does not move with soil, such as the frictional forces in the axial direction, (3) documentation of the behavior of buried structures during earthquakes, and (4) design methodologies.

Additional documentation of the behavior of buried structures in future earthquakes should provide more justification for the simplified analysis.

Appendix 1 - References

1. Archuleta, R. J., and Brune, J. N., 1975, "Surface Strong Motion Associated with a Stick-Slip Event in a Foam Rubber Model of Earthquakes," Bulletin of Seismological Society of America, vol. 65, pp. 1059-1071.

2. Goodling, E. C., Jr., 1978, "Flexibility Analysis of Buried Pipe," ASME/CSME Pressure Vessels and Piping Conference, Montreal, Canada, June 25-30.

3. Goodling, E. C., Jr., 1980, "More on Flexibility Analysis of Buried Pipe," ASME Pressure Vessels and Piping Conference, San Francisco, California, August 13-15.

4. Hanks, T. C., 1976, "Observations and Estimation of Long Period Strong Ground Motion in the Los Angeles Basin," International Journal of Earthquake Engineering and Structural Dynamics, vol. 4, pp. 473-488.

5. Katayama, T., 1976, "Effect of Ground Conditions of Seismic Damage to Buried Pipelines," Proceedings of U.S./Japan Seminar on Earthquake Engineering Research with Emphasis on Lifeline Systems, Tokyo, Japan.

6. Kubo, K., 1974, "Behavior of Underground Waterpipes During an Earthquake," Proceedings of the Fifth World Conference on Earthquake Engineering, June.

7. Lee, K. L., and Albaisa, A., 1974, "Earthquake Induced Settlements in Saturated Sands," Journal of the Geotechnical Engineering Division, ASCE, vol. 100, No. GT4, April, pp. 387-406.

8. Ligon, J. B., and Mayer, G. R., 1975, "Coefficient of Friction for Pipe Coating Materials," Pipe Line Industry, February.

9. Nakayama, S., Kiyomiya, O., and Tsuchida, H., 1977, "Observation of Dynamic Behavior of Kinuura Submerged Tunnel During Earthquakes," Proceedings of the Ninth Joint Meeting of the U.S./Japan on Wind and Seismic Effects, Tokyo, Japan.

10. Nasu, M., Kazama, S., Morioka, T., and Tamura, T., 1974, "Vibration Test of the Underground Pipe with a Comparatively Large Cross Section," Proceedings of the Fifth World Conference on Earthquake Engineering, June.

11. Newmark, N. M., 1967, "Problems in Wave Propagation in Soil and Rock," International Symposium on Wave Propagation and Dynamic Properties of Earth Materials, University of New Mexico Press.

12. Newmark, N. M. and Hall, W. J., 1978, "Development of Criteria for Seismic Review of Selected Nuclear Power Plants," NUREG/CR-0098, May.

13. Okamoto, S., Tamura, C., Kato, K., and Hamada, M., 1974, "Behaviors of Submerged Tunnels During Earthquakes," Proceedings of the Fifth World Conference on Earthquake Engineering, June.

14. O'Rourke, M. J., Singh, S., and Pikul, R., 1979, "Seismic Behavior of Buried Pipelines," Lifeline Earthquake Engineering - Buried Pipelines, Seismic Risk, and Instrumentation, ASME.

15. Parmelee, R. A., and Ludtke, C. A., 1975, "Seismic Soil Structure Interaction of Buried Pipelines," Proceedings of the U.S. National Conference on Earthquake Engineering.

16. Pyke, R., Seed, H. B., and Chan, C. K., 1975, "Settlement of Sands Under Multidirectional Shaking," Journal of the Geotechnical Engineering Division, ASCE, vol. 101, No. GT4, April, pp. 379-398.

17. Sakurai, A., and Takahashi, T., 1969, "Dynamic Stresses of Underground Pipe Lines During Earthquakes," Proceedings of the Fourth World Conference on Earthquake Engineering, Santiago, Chile.

18. Seed, H. B., and Silver, M. L., 1972, "Settlement of Dry Sands During Earthquakes," Journal of the Soil Mechanics and Foundations Division, ASCE, vol. 98, No. SM4, Proceedings Paper 8844, April, pp. 381-397.

19. Seed, H. B., 1976, "Evaluation of Soil Liquefaction Effects on Level Ground During Earthquakes," State-of-the-Art Paper Presented at the Symposium on Soil Liquefaction, ASCE National Convention, Philadelphia, October.

20. Shah, H. H., and Chu, S. L., 1974, "Seismic Analysis of Underground Structural Elements," Journal of the Power Division, ASCE, vol. 100, No. P01, July, pp. 53-62.

21. Shinozuka, M., and Kawakami, H., 1977, "Underground Pipe Damages and Ground Characteristics," Current State of Knowledge of Lifeline Earthquake Engineering, ASCE Specialty Conference.

22. Tamura, C., 1976, "Design of Underground Structures by Considering Ground Displacement During Earthquakes," Proceedings of the U.S./Japan Seminar on Earthquake Engineering Research with Emphasis on Lifeline Systems, Tokyo, Japan.

23. Tamura, C., Noguchi, T., and Kato, K., 1977, "Earthquake Observations Along Measuring Lines on the Surface of Alluvial Soft Ground," Proceedings of Sixth World Conference on Earthquake Engineering, New Delhi.

24. Tsuchida, H., and Kurata, E., 1976, "Observed Earthquake Ground Displacements Along a 2500 Meter Line," Proceedings of U.S./Japan Seminar on Earthquake Engineering Research with Emphasis on Lifeline Systems, Tokyo, Japan.

25. Yeh, G. K., 1977, "Seismic Analysis of Buried Metal or Concrete Pipes," ASME Third International Conference on Pressure Vessel Technology, Part 1 Analysis, Design and Inspection, Tokyo, Japan, April 19-22.

26. Youd, T. L., 1972, "Compaction of Sands by Repeated Shear Straining," Journal of the Soil Mechanics and Foundations Division, ASCE, vol. 98, No. SM7, Proceedings Paper 9-63, July, pp. 709-725.

Appendix 2 - Bibliography

Aisiks, E. G., and Tarshansky, I. W., "Soil Studies for Seismic Design of San Francisco Transbay Tube," Vibration Effects of Earthquakes on Soils and Foundations, ASTM STP 450, 1969, pp. 138-167.

Anderson, J. C., and Singh, A., "Seismic Response of Pipelines on Friction Support," Journal of the Engineering Mechanics Division, ASCE, vol. 102, No. EM2, April 1976, pp. 275-291.

Aoki, Y., and Hayashi, S., "Spectra for Earthquake Resistive Design of Underground Long Structures," Proceedings of the Fifth World Conference on Earthquake Engineering, vol. 1, 558, Rome, Italy, 1974, pp. 558-568.

Airman T., "A Review of the Earthquake Response and Aseismic Design of Underground Piping Systems," Proceedings of the Lifeline Earthquake Engineering Specialty Conference, ASCE, August 30-31, 1977, pp. 282-292.

"Structural Analysis and Design of Nuclear Plant Facilities," ASCE - Committee on Nuclear Structures and Materials of the Structural Division, New York, 1976.

Audibert, J.M.E., and Nyman, K. J., "Soil Restraint Against Horizontal Motion of Pipes," ASCE Geotechnical Engineering Division GT10, October 1977, pp. 1119-1141. Discussion by Akinmusura, J. O., GT9, September 1978; Thomas, H. P., GT9, September 1978; and Wang, L.R.L., GT12, December 1978.

Audibert, J.M.E., and Nymark, J., "Coefficients of Subgrade Reaction for the Design of Buried Piping," Structural Design of Nuclear Plant Facilities, Second Specialty Conference ASCE, New Orleans, Louisiana, December 8-10, 1975.

Ayres, J. M., Sun, T., and Brown, F. R., "Report on Damage to Building Mechanical Systems Due to the March 27, 1964, Alaska Earthquake," Ayres and Hayakawa, August 15, 1967.

Azar, J., Bomba, J., and Randolph, V., "Pipeline Stress Analysis," Transportation Engineering Journal, ASCE, vol. 101, No. TE1, February 1975, pp. 163-170.

Bealey, M., Lemons, J. D., Editors, "Concrete Pipe and Soil-Structure System," ASTM STP 630, Symposium presented at the 79th Annual Meeting, June 27 to July 2, 1976, ASTM, Chicago, Illinois, 1977.

Bechtel Power Corporation, "Seismic Analyses of Structures and Equipment for Nuclear Power Plants," Topical Report BC-TOP-4-A, September 1974, San Francisco, California.

Cajina, A., "The Managua Earthquake and Its Effects on Water Supply Systems," Proceedings of the Conference on Managua Earthquake, Earthquake Engineering Research Institute, vol. II, 1973, pp. 766-790.

Cast Iron Pipe Research Association, "Earthquakes and Other Geologic Hazards: Protective Design for Underground Pipe," Civil Engineering, ASCE, September 1976, p. 35.

Chang, P. Y., Hsu, D. J., and Pilkey, W., "Complex Infinite Beams on Elastic Foundation," Journal of Structural Division ASCE, ST 11, November 1977, pp. 2277-2282.

Christian, J. T., "Relative Motion of Two Surface Points During an Earthquake," Massachusetts Institute of Technology Research Report, R73-13, January 1973.

Davis, G. D., "Parametric Study of the Natural Frequencies and Mode Shapes for an Underground Piping System," Thesis presented to Rensselaer Polytechnic Institute at Troy, New York, in partial fulfillment of the requirements for the degree of Master of Science, 1977.

Design Criteria and Stress Analysis for the Trans-Alaska Pipeline, Summary Report, Alyeska Pipeline Service Company, December 18, 1972.

Dowding, C. H., "Lifeline Viability of Rock Tunnels-Empirical Correlations and Future Needs," The Current State of Knowledge of Lifeline Earthquake Engineering, ASCE, 1977, pp. 320-336.

Dowding, C. H., and Rozen, A., "Damage to Rock Tunnels from Earthquake Shaking," Proc. ASCE Journal, Geotechnical Engineering Division, GT2, February 1978, pp. 175-191.

Duke, C. M., "Damage to Water Supply Systems," The San Fernando, California Earthquake of February 9, 1971, Geological Survey Professional Paper 733, United States Geological Survey and National Oceanic and Atmospheric Association, 1971, pp. 225-240.

Duke, C. M., and Matthiesen, R. B., "Earthquakes Lifelines and ASCE," ASCE, Civil Engineering, vol. 43, December 1973, pp. 65-67.

Duke, C. M., and Moran, D. F., "Earthquakes and City Lifelines," San Fernando Earthquake of February 9, 1971, and Public Policy, Joint Committee on Seismic Safety of the California Legislature, 1972, pp. 53-67.

Duke, C. M., and Moran, D. F., "Guidelines for Evolution of Lifeline Earthquake Engineering," Proceedings of the United States National Conference on Earthquake Engineering, Earthquake Engineering Research Institute, 1975, pp. 367-376.

"Earthquake Damage Evaluation and Design Considerations for Underground Structures," A Public Service Paper, Los Angeles Section, ASCE, February 1974.

"Earthquake Damage to Water and Sewer Facilities," National Oceanic and Atmospheric Association/Earthquake Engineering Research Institute Subcommittee on Water and Sewerage Systems, Utilities, Transportation, and Sociological Aspects, C. M. Murphy, Editor, National Oceanic and Atmospheric and Administration, Washington, D.C., vol. 2, 1973, pp. 73-193.

"Earthquake and Other Geologic Hazards: Protective Design for Underground Pipe," Cast Iron Pipe Research Association, Civil Engineering, ASCE, vol. 46, No. 9, September 1976, p. 35.

Ford, D. B., "Design Considerations for Underground Pipelines in Geologically Hazardous Areas," Cast Iron Pipe News, Spring/Summer 1975.

Goodling, E. C., Jr., "Flexibility Analysis of Buried Pipe," ASME Paper 78-PVP-82, ASME/CSME Pressure Vessels and Piping Conference, Montreal, Canada, June 25-30, 1978.

Goodling, E. C., Jr., "More on Flexibility Analysis of Buried Pipe," ASME Pressure Vessels and Piping Conference, San Francisco, California, August 13-15, 1980.

Goto, H., Takada, S., and Nagao, S., "Effect of Joint Parts on Aseismic Strength of Buried Pipeline," Presented at the U.S./Japan Seminar on Earthquake Engineering Research with Emphasis on Lifeline Systems, Tokyo, Japan, November 1976.

Goto, H., Takada, S., and Nagao, S., "Effect of Joint Parts on Aseismic Strength of Buried Pipeline," Proceedings of the Sixth World Conference on Earthquake Engineering, pp. 12-105, New Delhi, India, 1977.

Goto, Y., Ota, J., and Takno, S., "On the Earthquake Response of Submerged Tunnels," Proceedings of the Fifth World Conference on Earthquake Engineering, vol. 1, 579, Rome, Italy, 1974.

Golze, A. R., "Use of Pipelines in the California Water Project," Journal of the Pipeline Division, ASCE, vol. 93, No. PL3, Proc. Paper 5584, November 1967, pp. 25-32.

Hadjian, A. H., Discussion on the paper "Earthquake Design Criteria for Subways," by R. T. Kuesel, Journal of the Structural Division, ASCE, vol. 96, 1970, pp. 159-160.

Haldar, A. K., Reddy, D. V., Arockiasamy, M., and Bobby, W., "Finite Element Nonlinear Seismic response Analysis of Submarine Pipe-Soil Interaction," Proceedings of International Symposium on Soils Under Cyclic and Transient Loading, vol. II, Swansea, U.K., January 1980.

Hall, W. J., and Newmark, N. M., "Seismic Design Criteria for Pipelines and Facilities," Journal of the Technical Councils of ASCE, vol. 104, No. TC1, November 1978, pp. 91-101.

Hamada, M., Izumi, H., and Omori, K., "Behavior of Underground Tanks During Earthquakes," Proceedings of the Fifth European Conference on Earthquake Engineering, vol. 1, chapter 3, Istanbul, Turkey, September 1975.

Hamada, M., Akimoto, T., and Izumi, H., "Dynamic Stress of a Submerged Tunnel During Earthquakes," Proceedings of the Sixth World Conference on Earthquake Engineering, New Delhi, India, January 1977.

Hemant, S., and Chu, S. L., "Seismic Analysis of Underground Structural Elements," Journal of Power Division, ASCE, vol. 100, No. PO1, Proc. Paper 10648, July 1974, pp. 53-62.

Hetenyi, M., Beams of Elastic Foundation, The University of Michigan Press, Ann Arbor, Michigan, 1946.

Howard, A. K., "Modulus of Soil Reaction (E') Values for Buried Flexible Pipe," Report No. REC-ERC-77-1, Bureau of Reclamation, Denver, Colorado, January 1977.

Howard, A. K., "Modulus of Soil Reaction Values for Buried Flexible Pipes," Journal Geotechnical Division, ASCE, GT1, January 1977, Discussion by Parmelle and Corotis, September, GT9; Nataraja, M. S., October, GT10; Spangler, M. G., October, GT10; and Smith, W. W., November, GT11.

Hrakilek, P. J., "Behavior of Underground Box Conduits in the San Fernando Earthquake of February 9, 1971," Technical Report E-72-1, Los Angeles Corps of Engineers, U.S. Army Engineer District, January 1972.

Hradilek, P. T., "Behavior of Underground Box Conduits in the San Fernando Earthquake," The Current State of Knowledge of Lifeline Earthquake Engineering, ASCE, 1977, pp. 308-319.

Iqbal, M. A., and Goodling, E. C., Jr., "Seismic Design of Buried Piping," 2d ASCE Specialty Conference on Structural Design of Nuclear Plant Facilities at New Orleans, Louisiana, December 8-10, 1975.

Isenberg, J., "The Role of Corrosion in the Seismic Performance of Buried Steel Pipes in Three United States Earthquakes," Weidlinger Associates, Consulting Engineers, Interim Grant Report No. IR-6 to NSF/RANN under Grant ENV P76-9838, March 1978.

Isenberg, J., et al, "Underground Pipelines in a Seismic Environment," Proceedings of the Lifeline Earthquake Engineering Specialty Conference, ASCE, August 30-31, 1977, pp. 267-281.

Iwan, W. D., "Earthquake Resistance of Public Utility Systems - A Report on Findings of the California Governor's Interagency Earthquake Committee," Proceedings of the U.S. National Conference on Earthquake Engineering, 1975, 387, EERI, Oakland, California.

Jennings, P. C., "Earthquake Problems of Networks and Systems," Proc. U.S./ Japan Seminar on Earthquake Engineering with Emphasis on Lifeline Systems, Tokyo, Japan, November 1976, pp. 5-14.

Johnson, J. A., "Possible Source Dependence of Ground Motion Along a Pipeline," Proceedings of the Sixth World Conference on Earthquake Engineering, New Delhi, India, January 1977, pp. 2-57 through 2-62.

Kachadoorian, R., "Earthquake: Correlation Between Pipeline Damage and Geologic Environment," Journal of AWWA, March 1976, pp. 165-167.

Kameswara, R.N.S.V., Das, Y. C., and Anandakrishnan, M., "Dynamic Response of Beams on Generalized Elastic Foundations," Int. J. Solids Structures, vol. 11, 1975, pp. 255-273.

Katayama, T., Kubo, K., and Sato, N., "Earthquake Damage to Water and Gas Distribution Systems," Proceedings of the U.S. National Conference on Earthquake Engineering 1975, Earthquake Engineering Research Institute, June 18-20, 1975, pp. 396-405.

Katayama, T., "Effect of Ground Conditions of Seismic Damage to Buried Pipelines," Proceedings of U.S./Japan Seminar on Earthquake Engineering Research with Emphasis on Lifeline Systems, U.S. National Science Foundation and Japan Society for the Promotion of Science, November 1976, pp. 197-210.

Katayama, T., Kubo, K., and Sato, N., "Quantitative Analysis of Seismic Damage to Buried Utility Pipelines," Proceedings of the Sixth World Conference on Earthquake Engineering, pp. 12-91, New Delhi, India, 1977.

Kennedy, R. P., Chow, A. W., and Williamson, R. A., "Fault Movement Effects on Buried Oil Pipeline," Transportation Engineering Journal of ASCE, vol. 103, No. TE5, September 1977, pp. 617-633.

Kennedy, R. P., et al, "General Considerations for Seismic Design of Oil Pipeline Systems," ASCE Lifeline Earthquake Engineering Specialty Conference, UCLA, Los Angeles, August 1977.

Kennedy, R. P., Chow, A. W., and Williamson, R. A., "Parametric Evaluation of Fault Movement Capacity of Buried Oil Pipeline," HN 5113.7R, Holmes and Narver, Incorporated, Orange, California, August 1976.

King, P. V., and Betz, J. M., "Earthquake Damage to a Sewer System," Journal of Water Pollution Control Federation, WPCF, May 1972, pp. 859-867.

Kitade, K., Shiraki, K., and Fukuzawa, K., "Aseismic Strength of Underground Pipelines," Technical Reports of Mitubishi Heavy Industries, vol. 11, No. 4, 1974, pp. 38-54.

Krizek, R. J., and McQuade, P. V., "Behavior of Buried Concrete Pipe," ASCE Journal, Geotechnical Division, July 1978.

Kubo, K., "Behavior of Underground Waterpipes During an Earthquake," Proceedings of the Fifth World Conference on Earthquake Engineering, vol. 1, June 1973, 569, Rome, Italy, 1974.

Kubo, K., "Fundamental Concept of Aseismic Design of Underground Piping Systems," Proceedings of the Fifth European Conference on Earthquake Engineering, vol. 2, chapter 6, Istanbul, Turkey, September 1975.

Kubo, K., Katayama, T., and Sato, N., "Quantitative Analysis of Seismic Damage to Buried Pipelines," Proceedings of the Fourth Japan Earthquake Engineering Symposium, Tokyo, Japan, 1975 (in Japanese).

Kubo, K., "Dynamic Behavior of Underground Pipes During Earthquake," SEISAN-KENKYU, Monthly Journal of Institute of Industrial Science, vol. 25, No. 2, University of Tokyo, Japan, February 1973 (in Japanese).

Kubo, K., and Katayama, T., "Survey on Underground Pipe Damages Due to Earthquake," Unpublished Report, Tokyo Metropolitan Disaster Prevention Congress, Governor's Office, The Metropolitan Government of Tokyo, Japan (in Japanese).

Kuesel, R. T., "Earthquake Design Criteria for Subways," Journal of the Structural Division, ASCE, vol. 95, No. ST6, June 1969, pp. 1213-1231.

Kuribayashi, E., Iwasaki, T., and Kawashima, K., "Dynamic Behavior of a Subsurface Tubular Structure," Fifth Symposium on Earthquake Engineering, Roorkee, India, July 1974.

Kuribayashi, E., Iwasaki, T., and Kawashima, K., "Dynamic Behavior of a Subsurface Tubular Structure," Proceedings of the Fifth European Conference on Earthquake Engineering, vol. 1, chapter 3, Istanbul, Turkey, September 1975.

Kurihara, C., and Sakurai, A., "A Method of Stress Reduction for Thermal and Earthquake Condition of Underground Pipelines," Technical Report No. 72005, C.R.I.E.P.I., September 1972.

Luk, N. N., "Dynamic Response of a Simple Piping System on an Elastic Foundation," Master of Engineering Project Report, Department of Civil Engineering, Rensselaer Polytechnic Institute, Troy, New York, September 1976.

Luscher, U., Black, W. T., and Nair, K., "Geotechnical Aspects of Trans-Alaska Pipeline," Transportation Engineering Journal, ASCE, vol. 101, No. TE4, Proc. Paper 11722, November 1975, pp. 669-680.

Miyahara, F., and Ergatoudis, J. C., "Matrix Analysis of Structure-Foundation," Journal of the Structural Division, ASCE, vol. 102, No. ST1, January 1976, pp. 251-265, Discussion: B. Deschapelles, ST12: December 1976, Closure ST8: August 77.

Miyajima, N., et al, "An Example of Seismic Design and Earthquake Response Measurement of a Buried Pipeline," U.S./Japan Seminar on Earthquake Engineering Research with Emphasis on Lifeline Systems, U.S. National Science Foundation and Japan Society for the Promotion of Science, November 1976.

Miyamoto, H., et al, "Experiments on the Behavior of Underground Pipes During Earthquakes," Proceedings of the Third Japan Earthquake Engineering Symposium, November 1970, pp. 129-132.

Muto, K., Ushida, K., and Tsugama, T., "The Earthquake Response Analysis of Underground Tunnels," Proceedings of the Third Japan Earthquake Engineering Symposium, 437, November 1970.

Nakano, Y., et al, "A Consideration on the Seismic Response Analysis of Underground Structures," Technical Report on Mitubishi Heavy Industries, 1974.

Nakayama, S., Kiyomiya, O., and Tsuchida, H., "Observation of Dymanic Behavior of Kinuura Submerged Tunnel During Earthquakes," Presented at the May 1977 Ninth Joint Meeting of the United States and Japan on Wind and Seismic Effects, Tokyo, Japan.

Narita, K., "Effect of Earthquakes on Pipelines," High Pressure, 1971 (in Japanese).

Narita, K., "An Opinion on Structural Reliability of Pipelines," presented at the Japan/U.S. Joint Seminar on "Reliability Approach in Structural Engineering," Tokyo, Japan, May 1974.

Narita, K., "Study on Pipeline Failure Due to Earthquake," presented at the Japan/U.S. Joint Seminar on Lifeline Systems, Tokyo, Japan, November 1976.

Nasu, N., et al, "Vibration Test of the Underground Pipe with a Comparatively Large Cross-Section," Proceedings of the Fifth World Conference on Earthquake Engineering, vol. 1, 583, Rome, Italy, 1974.

Nath, P., "Finite Element Analysis of a Large Diameter Buried Steel Pipeline," TRRL-LR-778, Transport and Road Research Laboratory, Crowthorne, England, 1977.

Newmark, N. M., "Problems in Wave Propagation in Soil and Rock," Proceedings of the International Symposium on Wave Propagation and Dynamic Properties of Earth Materials, Albuquerque, New Mexico, 1968.

Newmark, N. M., and Hall, W. J., "Seismic Design Criteria for Nuclear Reactor Facilities," Proceedings of Fourth World Conference on Earthquake Engineering, Santiago, Chile, 1969.

Newmark, N. M., and Hall, W. J., "Pipeline Design to Resist Large Fault Displacement," Proceedings of the U.S. National Conference on Earthquake Engineering, 1975, EERI, Oakland, California, pp. 416-425.

Novak, M., and Hindy, A., "Seismic Response of Buried Pipelines," Second Annual ASCE Engineering Mechanics Division, Specialty Conference, May 23-25, 1977, pp. 44-50.

Okamoto, S., et al, "Behaviors of Submerged Tunnels Survey Earthquakes," Proceedings of the Fifth World Conference on Earthquake Engineering, vol. 1, 544, Rome, Italy, 1974.

Okamoto, S., and Tamura, C., "Behavior of Subaqueous Tunnel During Earthquakes," International Journal of Earthquake Engineering and Structural Dynamics, John Wiley & Sons, Incorporated, New York, New York, vol. 1, No. 3, January-March, 1973, pp. 253-266.

O'Rourke, M., and Wang, L.R.L., "Earthquake Response of Buried Pipeline," presented at the June 19-21, 1978, ASCE Geotechnical Engineering Division Specialty Conference, Pasadena, California.

O'Rourke, M. J., and Wang, L.R.L., "Seismic Shaking of Buried Pipelines," ASCE Annual Convention, Chicago, Illinois, October 1978.

Panoussis, G., "Seismic Reliability of Lifeline Networks," SDDA Report No. 15, MIT Department of Civil Engineering Report, R74-57.

Parmelke, R. A., and Corotis, R. B., "Modulus of Soil Reaction Values for Buried Flexible Pipe," ASCE - Geotechnical Division, GT9, September 1977, pp. 1034-1036.

Parmelee, R. A., and Ludtke, C. A., "Seismic Soil-Structure Interaction of Buried Pipelines," Proceedings of the U.S. National Conference on Earthquake Engineering 1975, Earthquake Engineering Research Institute, June 18-20, 1975, pp. 406-415.

Parnes, R., "Dynamic Response of Imbedded Cylindrical Shell," Journal of the Engineering Mechanics Division, ASCE, vol. 97, No. EM2, Proc. Paper 8073, April 1971, pp. 421-440.

Patel, Y. A., Cho, F. L., and Dimopoulos, A. P., "Pipeline Design for Earthquake Zones," Pipeline Gas Journal, vol. 202, No. 14, December 1975.

Pearson, F. H., "Beam Behavior of Buried Rigid Pipelines," Journal of the Environmental Engineering Division, ASCE, vol. 103, No. EE5, October 1977, pp. 761-783.

Rashidov, T., "The Principles of the Dynamic Theory of Earthquake Resistance of the Complex Systems of Underground Structures," Proceedings of the Fifth World Conference on Earthquake Engineering, vol. 1, 593, Rome, Italy, 1974.

"Research on Earthquake Protection of Petroleum Pipelines," Japan Road Association, March 1974 (in Japanese).

Richardson, C. B., "Damage to Utilities," The Great Alaska Earthquake of 1964 - Engineering, The Committee of the Alaska Earthquake, National Academy of Science, Washington, D.C., 1973, pp. 1034-1073.

Sakurai, A., and Takahashi, T., "Dynamic Stresses of Underground Pipe Lines During Earthquakes," Proceedings of the Fourth World Conference on Earthquake Engineering, Santiago, Chile, 1969.

Sakurai, A., "Earthquake Resistivity of Underground Pipelines," Reprint of Regional Symposium of JSSMFE, Hokkaido Branch, October 1970.

Sakurai, A., et al, "Dynamic Stresses of Underground Pipelines During Earthquakes, Rept. No. 2," Technical Report No. 69087, C.R.I.E.P.I., April 1970.

Sakurai, A., et al, "A Proposal for Earthquake Response Analysis of Long Structures," Proceedings of the Third Japan Earthquake Engineering Symposium, November 1970.

Sakurai, A., Kurihara, C., and Yazima, H., "Earthquake Response Analysis Method of Long Structures as Multi-Input System," Technical Report No. 72004, C.R.I.E.P.I., September 1972.

Sakurai, A., Durihara, C., and Takahashi, T., "A Proposal for Earthquake Response Analysis of Long Structures and Aseismic Design Criteria of Pipe Lines," Proceedings of the Fifth World Conference on Earthquake Engineering, vol. 1, 603, Rome, Italy, 1974.

Schrock, B. J., "Installation of Fiberglass Pipe," Transportation Engineering Journal, ASCE, vol. 104, No. TE6, November 1978, pp. 835-846.

Selig, E. T., et al, "Long-Span Buried Structure Design and Construction," ASCE Journal Geotechnical Engineering Division, GT10, October 1978, pp. 953-967.

Selig, E. T., and Chang, C. S., "Measured and Predicted Response of Buried Flexible Structure," presented at ASCE Annual Convention, October 16-20, 1978, Chicago, Illinois.

Selig, E. T., Abel, J. F., Kudhawy, F. H., and Falby, W. E., "Long-Span Buried Structure Design and Construction," ASCE Journal, Geotechnical Engineering Division, GTf, July 1978, pp. 953-966.

Shah, H. H., and Chu, S. L., "Seismic Analysis of Underground Structural Elements," Journal of the Power Division of ASCE, vol. 100, No. PO1, July 1974, pp. 53-62.

Shinozuka, M., and Kawakami, H., "Underground Pipe Damages and Ground Characteristics," Proceedings of the Lifeline Earthquake Specialty Conference, ASCE, August 30-31, 1977, pp. 239-307.

Sugimura, Y., "Seismic Strain Induced in the Ground During Earthquakes," UCB/EERC-74/14, University of California, Richmond Earthquake Engineering Research Center, June 1977, p. 73.

Takada, S., "Earthquake Resistant Design of Underground Pipelines," Proceedings of the Sixth World Conference on Earthquake Engineering, New Delhi, India, January 1977, pp. 12-97 through 12-102.

Takada, S., and Nagao, S., "Efficiency of Joint Parts for Aseismic Strength of Buried Pipelines," Proceedings of the Fourth Japan Earthquake Engineering Symposium, 1975, pp. 679-686.

Takada, S., and Nagao, S., "Dynamic Frictional Forces and Efficiency of Joint Parts for Aseismic Strength of Buried Pipelines," Proceedings of U.S./Japan Seminar on Earthquake Engineering Research with Emphasis on Lifeline Systems, U.S. National Science Foundation and Japan Society for the Promotion of Science, November 1976, pp. 211-222.

Takada, S., and Wright, J. P., "Dynamic Soil Stiffness, Mass and Damping Coefficients for Buried Pipelines," NSF Research Memorandum RM-2, Weidlinger Associates, September 1977.

Taleb-Agha, G., "Seismic Risk Analysis of Lifeline Networks," Bulletin Seismological Society of America, vol. 67, No. 6, December 1977, pp. 1625-1645.

Taleb-Agha, G., and Whitman, R. V., "Seismic Risk Analysis of Discrete Systems of Sites," Earthquake Engineering and Structural Dynamics, vol. 5, 1977, pp. 293-304.

Tamura, C., "Design of Underground Structures by Considering Ground Displacement During Earthquakes," Proceedings of U.S./Japan Seminar on Earthquake Engineering Research with Emphasis on Lifeline Systems, Tokyo, Japan, November 1976, pp. 417-434.

Tamura, C., and Okamoto, S., "A Study on the Earthquake Resistant Design of Subaqueous Tunnels," Proceedings of the Fifth European Conference on Earthquake Engineering, Istanbul, Turkey, 1975.

Tamura, C., and Okamoto, S., "On Earthquake Resistant Design of a Submerged Tunnel," Proceedings of the International Symposium on Earthquake Structural Engineering, University of Missouri, Columbia, Missouri, August 1976, pp. 809-822.

Tamura, C., Okamoto, S., and Hamada, M., "Dynamic Behavior of a Submerged Tunnel During Earthquakes," Report of the Institute of Industrial Science, Serial No. 154, vol. 24, No. 5, University of Tokyo, Tokyo, Japan, March 1975.

Tamura, C., Okamoto, S., and Kato, K., "Observations on Dynamic Strains of Submerged Tunnel During Earthquakes," Bulletin of Earthquake Resistant Structure Research Center, University of Tokyo, Tokyo, Japan, No. 6, December 1972, pp. 1-21.

Tamura, C., and Okazaki, T., "Dynamic Model Test of French Type Tunnel," Reprint of the Ninth Earthquake Engineering Symposium on Civil Engineering Structures, July 29, 1972.

"Technical Guidance for Petroleum Pipeline," Japan Road Association, March 1974 (in Japanese).

Terzaghi, K., "Evaluation of Coefficient, of Subgrade Reaction," Geotechnique, vol. 5, 1955, pp. 297-326.

Toki, K., and Takada, S., "Earthquake Response Analysis of Underground Tubular Structure," Bulletin of the Disaster Prevention Research Institute, Kyoto University, Kyoto, Japan, vol. 24, part 2, June 1974.

Tsuchida, H., and Kurata, E., "Observed Earthquake Ground Displacements Along a 2,500 Meter Line," Proceedings of U.S./Japan Seminar on Earthquake Engineering Research with Emphasis on Lifeline Systems, Tokyo, Japan, November 1976, pp. 29-42.

Wang, L.R.L., "Vibration Frequencies of Buried Pipelines," Journal of the Technical Councils of ASCE, vol. 104, No. TC1, November 1978, pp. 71-81.

Wang, L.R.L., "Overview of Buried Pipelines Under Seismic Loading," Journal of the Technical Councils of ASCE, vol. 104, No. TC1, November 1978, pp. 121-130.

Wang, L.R.L., and Cheng, K. M., "Seismic Response Behavior of Buried Pipelines," ASME Annual Winter Convention, San Francisco, California, December 11-15, 1978.

Wang, L.R.L., and Fung, R.C.Y., "Seismic Design Criteria for Buried Pipelines," ASCE Pipeline Division Specialty Conference, New Orleans, Louisiana, January 1978.

Wang, L.R.L., and O'Rourke, M. J., "State-of-the-Art of Buried Lifeline Earthquake Engineering," Proceedings of the Lifeline Earthquake Engineering Specialty Conference, ASCE, August 30-31, 1977, pp. 252-266.

Weidlinger, P., "Behaviour of Underground Lifelines in Seismic Environment," Report IR-4, Weidlinger Associates, New York, New York, 1977, p. 22.

Werner, S. D., Lee, L. C., Wong, H. L., and Trifunac, M. D., "An Evaluation of the Effects of Traveling Seismic Waves on the 3-D Response of Structures," R-7720-4514, Agbabian Associates, El Segundo, California, October 1977. Also, ASCE Convention and Exposition, Chicago, Illinois, October 16-20, 1978.

Whitman, R. V., Cornell, C. A., and Taleb-Agha, G., "Analysis of Earthquake Risk for Lifeline Systems," Proceedings of the United States National Conference on Earthquake Engineering, Earthquake Engineering Research Institute, 1975, pp. 377-386.

Yamada, T., and Ogihara, Y., "Seismic Response Analysis of Underground Pipeline for Emergency Core Cooling System in Nuclear Power Plant," Paper K7/4, vol. IV, Second SMIRT Conference, Berlin, Germany, pp. 14, 1973.

Yeh, G. K., "Seismic Analysis of Buried Metal or Concrete Pipes," ASME Third International Conference on Pressure Vessel Technology, Part 1 Analysis, Design and Inspection, Tokyo, Japan, April 19-22, 1977, pp. 389-397.

Yeh, G.C.K., "Seismic Analysis of Long Buried Structures," Second International Conference on Structural Mechanics in Reactor Technology, Paper K4/8, vol. IV, Berlin, Germany, 1973.

Yeh, G. K., "Seismic Analysis of Slender Buried Beams," Bulletin of Seismological Society of America, vol. 64, No. 5, October 1974, pp. 1551-1562.

Zaslawsky, M., Blake, A., and Karpenko, V. N., "Seismic Safety Considerations for Buried Pipes," UCRL-76977, Lawrence Livermore Laboratory, Livermore, California, July 18, 1975.

APPENDIX 3. - SEISMIC RESPONSE OF BURIED STEEL PIPE

PROBLEM STATEMENT:

Determine the seismically induced loads and displacement for the given piping system.

GIVEN DATA:

The pipe layout is shown on page 37. Pipe data and soil data are given on page 38. The piping system shall be designed for a near-field earthquake with a maximum ground acceleration of 0.25g coupled with a NRC regulatory guide 1.60 response spectra. The propagation velocity of the seismic waves is assumed to be 2500 fps (762.5 m/sec) and the maximum ground velocity is 12 in/sec (30.5 cm/sec).

PIPE LAYOUT:

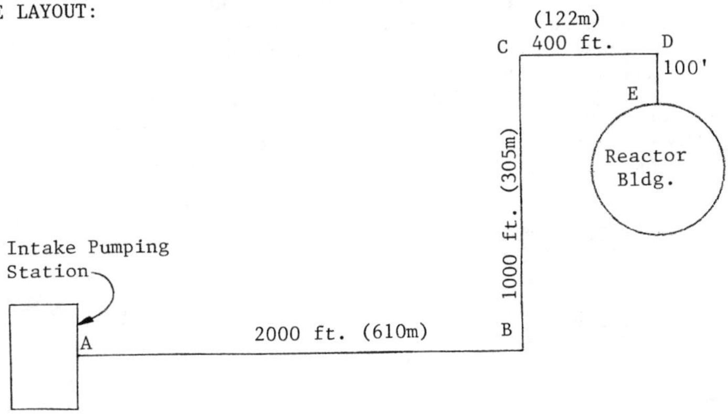

Pipe Data	Soil Data
Carbon Steel SA 106 Grade B	Uniform depth of soil Depth = H = 30 ft (9.2 m) Pipe Depth = 10 ft (3.1 m)
$t = 0.375$ in. (0.952 cm) ID = 29.25 in. (74.3 cm) OD = 30.00 in. (76.2 cm) Elbows are long radius	Shear wave velocity = 700 fps (213 m/sec) Comp. wave velocity = 1200 fps (366 m/sec) Poisson's Ratio, $\mu = 0.30$
$A = 34.9\ in^2\ (225.2\ cm^2)$ $I_p = 3823\ in^4\ (159125.3\ cm^4)$ $Z = 254.8\ in^3\ (4175.4\ cm^3)$	Shear Modulus = G_s $= 1.9 \times 10^6$ ksf (920500 kg/cm^2) Young's Modulus = E_s = 4.9×10^6 ksf (2,374,000 kg/cm^2)
Allowable Stress = 15,000 psi (1046 kg/cm^2) Yield Stress = 35,000 psi (2442 kg/cm^2) Min. Ult. Stress = 60,000 psi (4186 kg/cm^2) Young's Modulus = E_p = 28,000,000 psi (1,953,488 kg/cm^2)	Cohesion = 400 psf (0.2 kg/cm^2) Friction angle = $\phi = 15°$ Saturated Weight = 125 pcf (2 g/cm^3) Moist Weight = 120 pcf (1.9 g/cm^3) Water table = 15 ft. (4.6 m) down

LOADS ON STRAIGHT SECTIONS OF PIPE

Axial:

$$\varepsilon_a = \pm \frac{V}{C} = \frac{12}{2500 \times 12} = 4 \times 10^{-4} \text{ in/in}$$

$$F_a = \varepsilon AE = 0.004(34.9)(2.8 \times 10^7)$$

$$F_a = 391000 \text{ lbs } (175950 \text{ kg.})$$

Using the empirical relationship of Sakurai and Takahashi, the maximum axial load from the axial strain cannot exceed

$$F_{a_{max}} = f\ L/4$$

where f = friction force per unit length of pipe
L = wave length of seismic waves

The friction force per unit length of the pipe is the shear transfer capacity between the steel pipe and the soil. Assume this friction coefficient is 0.5. At 10 feet down

$\sigma = \gamma h = 120(10) = 1200$ psf (0.6 kg/cm^2)

$f = 1200(0.5)\pi(2.5) = 4710$ lb/ft. (69.5 kg/cm)

Also check soil-soil failure just away from the pipe. The shear stress is

τ = cohesion + $\sigma \tan \phi$

$\tau = 400 + 1200 \tan 15° = 720$ psf (0.35 kg/cm^2)

In using this approach, no distinction is made for σ being due to vertical or horizontal which is conservative.

$f = \pi\ (2.5)\ 720 = 5700$ lb/ft. (84.2 kg/cm)

Therefore use $f = 4710$ lb/ft. (69.5 kg/cm)

$L = Tc$ (wave length)

where T is the natural period of the seismic waves. The predominant natural period of the waves can be determined from a site specific analysis. For this case using a regulatory guide spectrum assume a predominant period of 0.2 second.

$L = 0.2\ (2500) = 500$ ft. (152.5 m)

Therefore

$F_{a_{max}} = 4710\ (500/4) = 589{,}000$ lbs (265050 kg)

which is greater than 391,000 lbs so use

$F_a = 391{,}000$ lbs (175950 kg)

Bending:

$$\varepsilon_b = \pm \frac{R_a}{C^2} = \frac{15\ (96.6)}{(2500 \times 12)^2} = 1.6 \times 10^{-6} \text{ in/in}$$

$\sigma_b = \varepsilon E = 1.6 \times 10^{-6}\ (28 \times 10^6) = 45$ psi (3.1 kg/cm^2)

$M_b = \sigma_b\ Z = 45(254.8) = 11500$ in-lb (13144 cm-kg)

LOADS AT THE ELBOWS

The loads at the elbows are determined using the methods proposed by Goodling (references 2 and 3). For this example problem only the loads at elbow B will be calculated.

Intermediate Parameters:

Eq. 9 (ref. 2)

$$k_o = \frac{0.65}{D}\left[\frac{E_s D^4}{E_p I_p}\right]^{1/12} \frac{E_s}{1 - \mu^2}$$

$$k_o = \frac{0.65}{30}\left[\frac{34000(30)^4}{28 \times 10^6(3823)}\right]^{1/12} \frac{34000}{1 - 0.3^2}$$

$$= 730 \text{ pci } (20 \text{ kg/cm}^3)$$

$$f = 4710 \text{ lbs/ft} = 392 \text{ lbs/in } (69.4 \text{ kg/cm})$$

Eq. 11 (ref. 2)

$$k = k_o D = 730(30) = 21900 \text{ psi } (1528 \text{ kg/cm}^2)$$

Eq. 12 (ref. 2)

$$\lambda = \sqrt[4]{\frac{k}{4EI}} = \left[\frac{21900}{4(28 \times 10^6)(3823)}\right]^{1/4}$$

$$= 0.015 \text{ in}^{-1} \ (0.006 \text{ cm}^{-1})$$

Eq. 13 (ref. 2)

$$\ell_m = \frac{\varepsilon AE}{f} = \frac{F_a}{f} = \frac{39100}{4710} = 83 \text{ ft } (25.3 \text{ m})$$

Eq. 14 (ref. 2)

$$L' = \frac{4}{3}\Omega\left[\sqrt{1 + \frac{3 1m}{2\Omega}} - 1\right]$$

$$\Omega = \frac{AE\lambda}{k} = \frac{34.9(28 \times 10^6)(0.015)}{21900} = 670 \text{ in. } (17 \text{ m})$$

$$L' = \frac{4}{3}(670)\left[\sqrt{1 + \frac{3(996)}{2(670)}} - 1\right] = 712 \text{ in} = 59 \text{ ft } (18 \text{ m})$$

Classification of Elbow:

Looking down leg AB at B -

ℓ_2 = 2000 ft = 24000 in (610 m) (P leg) > L' $\Longrightarrow$ "long"

ℓ_1 = 1000 ft = 12000 in (305 m) (T leg) > $\frac{3\pi}{4\lambda} \Longrightarrow$ "long"

therefore this is Type 3 - LPLT

Looking down leg CB at B -

ℓ_2 = 1000 ft = 12000 in (305 m) (P leg) > $L^1 \Longrightarrow$ "long"

ℓ_1 = 2000 ft = 24000 in (610 m) (T leg) > $\frac{3\pi}{4\lambda} \Longrightarrow$ "long"

therefore this is Type 2 - LPLT

Reference 2 by Goodling considers the longitudinal and transverse legs flexible with the elbow inflexible. Reference 3 by Goodling considers both legs and elbow as flexible. Therefore determine loads from both approaches for comparison.

Elbow Inflexible:

Eq. 16 (ref. 2)

$$S = \varepsilon AE - fL'$$
$$S = 391000 - 4710\ (59) = 113000 \text{ lbs } (50850 \text{ kg})$$

Eq. 17 (ref. 2)

$$M = \frac{S}{3\lambda} = \frac{113000}{3(0.015)} = 2{,}510{,}000 \text{ in-lb } (2868930 \text{ cm-kg})$$

Eq. 18 (ref. 2)

$$\Delta = \frac{4\lambda S}{3k} = \frac{4(0.015)(113000)}{3(21{,}900)} = 0.1 \text{ in. } (0.25 \text{ cm})$$

These forces are the same for both conditions, namely elbow B looking down leg AB or leg CB.

Elbow Flexible:

(ref. 3) $$a_{11} = \varepsilon L' - \frac{fL'^2}{2AE}$$

$$a_{11} = 4 \times 10^{-4}(712) - \frac{392(712)^2}{2(34.9)(28 \times 10^6)} = 0.183$$

$$a_{12} = -\frac{L'}{AE} = -\frac{712}{34.9(28 \times 10^6)} = -7.3 \times 10^{-7}$$

$$a_{21} = \frac{\lambda}{r\phi} KEI_p$$

where r = elbow radius = 45 in (114.3 cm)

ϕ = elbow angle = 90° = 1.57 radians

$$K = 1 - \frac{9}{10 + 12\left(\frac{tr}{r_o^2}\right)^2}$$

in which t = wall thickness = 0.375 in (0.95 cm)

r_o = pipe section outside radius
= 15 in. (38.1 cm)

$$K = 1 - \frac{9}{10 + 12\left(\frac{.375 \times 45}{15^2}\right)^2}$$

$$K = 0.11$$

$$a_{21} = \frac{0.015}{45(1.57)} (0.11)(28 \times 10^6)(3823) = 2.5 \times 10^6$$

$$a_{31} = \frac{k}{2\lambda} = \frac{21900}{2(0.015)} = 730000$$

$$a_{32} = \lambda = 0.015$$

Eq. 11 (ref. 3)

$$\Delta_1 = \frac{a_{11}}{1 - a_{12}a_{31} - a_{12}a_{21}a_{32}}$$

$$\Delta_1 = 0.12 \text{ in}$$

Eq. 8a (ref. 3)

$$M = a_{21}\Delta_1 = 2.5 \times 10^6(0.12) = 300{,}000 \text{ in-lb} \quad (342900 \text{ cm-kg})$$

Eq. 9a (ref. 3)

$$S_1 = a_{31}\Delta_1 + a_{32}M$$

$$S_1 = 730{,}000(0.12) + 0.015(300{,}000) = 92100 \text{ lbs} \quad (41445 \text{ kg})$$

ASME Code Allowable Stresses

The ASME code does not address buried piping. Therefore the code is not directly applicable to our problem, however for this example it will be used two ways. The first way will consider the earthquake stresses as primary and the second way as secondary.

Primary:

The ASME equations to be satisfied are as follows:

1. Sustained loads

$$S_{SL} = \frac{PD_o}{4t_n} + \frac{0.75i\ M_a}{Z} \leq 1.0S_h$$

P = internal design pressure, psi

D_o = outside diameter of pipe, in.

t_n = nominal wall thickness, in.

M_a = resultant moment loading on cross section due to weight and other sustained loads, in-lb.

Z = section modulus of pipe, in^3

i = stress intensification factor

S_h = basic material allowable stress for this case is 15,000 psi (1046 kg/cm^2)

i = 5.0 for this example (intensification factor)

2. Occasional loads

$$S_{OL} = \frac{P_{max} D_o}{4t_n} + 0.75i \left(\frac{M_A + M_B}{Z} \right) + \frac{S}{A} \leqq 1.2S_h$$

P_{max} = peak pressure, psi

M_B = resultant moment loading on cross section due to occasional loads, such as thrusts from relief and safety valve loads from pressure and flow transients, and earthquake.

$\frac{S}{A}$ = modification to ASME equation to account for earthquake axial stress

For buried piping M_B is just the earthquake loading. The allowable $1.2S_h$ is for the OBE and $1.8S_h$ would correspond to the SSE emergency case.

3. Thermal expansion

The requirements of either of the following equations must be met

$$S_E = i \frac{M_c}{Z} + \frac{S}{A} \leqq S_A$$

M_c = range of resultants moments due to thermal expansion

S_A = allowable stress range for this case is $1.5S_h$
= 22,500 (1570 kg/cm^2)

$\frac{S}{A}$ = modification for thermal axial load

$$S_{TE} = \frac{PD_o}{4t_n} + 0.75 \frac{M_A}{Z} + i \frac{M_c}{Z} + \frac{S}{A} \leqq S_h + S_A$$

For most buried pipes which are low pressure raw water pipes, the design pressure term plus the soil overburden pressure can be neglected. Also for most buried pipes continuously supported by soil the M_A term can be neglected. Therefore the SSE emergency case would control.

Check the stress in the elbow for the SSE considering the elbow flexible.

$$S_{OL} = \frac{0.75i\ M_B}{Z} + \frac{S}{A} \leqq 1.8S_h = 27000 \text{ psi } (1884 \text{ kg/cm}^2)$$

$$S_{OL} = \frac{0.75(5)(300{,}000)}{254.8} + \frac{92{,}100}{34.9}$$

$$S_{OL} = 7{,}050 \text{ psi } (492 \text{ kg/cm}^2) \leqq 27000 \text{ psi } (1884 \text{ kg/cm}^2) \text{ O.K.}$$

Assuming a 65°F change in temperature in the pipe the thermal moment and axial load for a flexible elbow is

M_C = 230,000 in-lb. (262890 cm-kg)

S = 25000 lbs (11250 kg)

$$S_E = i\frac{M_c}{Z} + \frac{S}{A} = 5230 \text{ psi } (365 \text{ kg/cm}^2) < 22{,}500 \text{ psi } (1570 \text{ kg/cm}^2) \text{ O.K.}$$

Secondary:

The ASME code allows the effect of earthquake anchor movements to be considered as secondary stresses. The earthquake induced displacements on buried pipes are similar to the anchor movements. The soil is the driving force on the pipe. The soil will experience certain deformations due to the earthquake. These deformations are the maximum deformation that may be imposed on the pipe. The pipe may not deform beyond the soil imposed deformations. Thus the pipe behavior is limited. Stresses thus induced cannot grow without bound. Therefore they are in essence self-limiting. This is the basic characteristic of a secondary stress. When the earthquake stresses are considered as secondary the following ASME equations are applicable. The code does not require secondary stresses to be checked for the emergency or faulted conditions. Therefore the equations are really only applicable for the OBE case, but for this example they will be used for the SSE case. If the earthquake stresses are going to be considered as secondary, then the ASME code should be revised for buried piping.

$$S_E = i\frac{M_c}{Z} + \frac{S}{A} \leqq S_A$$

M_c = resultant moment from thermal expansion plus earthquake

$\frac{S}{A}$ = modification to equation to account for axial stress from earthquake and thermal expansion.

or

$$S_{TE} = \frac{PD_o}{4t_n} + 0.75i\,\frac{M_A}{Z} + i\,\frac{M_c}{Z} + \frac{S}{A} \leqq S_h + S_A$$

Check the stress considering the elbow flexible. The ASME code treats secondary stresses in such a manner that the stress range rather than the maximum stress must be within the allowable stress S_A. For a buried pipe, the secondary stress should be taken as the worst combination of thermal and seismic stresses. The maximum secondary stress is the sum of the thermal stress and the positive seismic stress, and the minimum stress is the sum of the thermal stress and the negative seismic stress. If the thermal stress exceeds the peak seismic stress, then the secondary stress range can be expressed by $\sigma_{thermal} + \sigma_{seismic}$. If the thermal stress is less than the peak seismic stress, then the secondary stress range is equal to twice the peak seismic stress.

Calculate maximum secondary stress range:

Thermal stress = 5230 psi (365 kg/cm^2)

$$\text{Peak seismic stress} = \frac{5.0(300{,}000)}{254.8} + \frac{92{,}100}{34.9} = 8525 \text{ psi } (595 \text{ kg/cm}^2)$$

Since seismic stress is greater than the thermal, then the secondary stress range is

$$S_R = 2(\text{Seismic Stress}) = 17{,}050 \text{ psi } (1190 \text{ kg/cm}^2)$$

which is less than 22500 psi (1570 kg/cm^2), therefore there is no reason to check S_{TE}.

For this example problem the stresses are within the allowables, therefore the pipe is O.K. In some cases, the allowable stresses may be exceeded, requiring modifications. If the elbow had been considered inflexible for this example, the allowable stresses would have been exceeded. One way to modify is to use flexible couplings. When flexible couplings are incorporated in the design, it is necessary to calculate the maximum conceivable seismic movement of the soil relative to the pipe ends so that the pipe gap and a sufficiently long middle ring or bellows assembly can be specified. This movement can be expressed by:

$$\delta = 2(\varepsilon_m \ell_m - f\ell_m^{\,2}/2AE)$$

where $\varepsilon_m \ell_m$ is the total deformation of the soil at each free end over the minimum slippage length ℓ_m and $f\ell_m^{\,2}/2AE$ is the amount of pipe deformation over the same length. For this example

$\varepsilon_m = 4 \times 10^{-4}$ in/in

$\ell_m = 996$ in (25.3 m)

$f = 4710$ lb/ft/12 = 392 lb/in (69.5 kg/cm)

$$\delta = 2 \left[4 \times 10^{-4}(996) - \frac{392(996)^2}{2(34.9)(28 \times 10^6)} \right]$$

$\delta = 0.4$ inch (1.0 cm)

Therefore a pipe end gap of twice this value or about 1.0 inch (2.54 cm) should be specified.

APPENDIX 4. - SEISMIC RESPONSE OF BURIED CONCRETE ELECTRICAL DUCT BANK

PROBLEM STATEMENT:

Determine the seismically induced loads and displacements for the given electrical conduit bank.

GIVEN DATA:

The conduit cross-section, conduit data, and soil data are given on page 49. The conduit shall be designed for a near-field earthquake with a maximum ground acceleration of 0.25g coupled with a NRC regulatory guide 1.60 response spectra. The propagation velocity of the seismic waves is assumed to be 2500 (762.5 m/sec) fps and the maximum ground velocity is 12 in/sec (30.5 cm/sec).

DISCUSSION:

As indicated in the text, the design methodology for buried concrete structures depends a great deal on the function of the structure. If considerable cracking can be allowed and the purpose of the rebar is simply to hold the concrete together, then the concrete compressive strain should be limited to an ultimate of 0.3 percent. This is a straight forward check unless there are other loads in addition to earthquake on the structure, or if the function of the structure requires it to maintain the integrity of its boundaries. The following example gives a brief way to calculate "psuedo strain dependent" loads which can be combined with other loads in a normal working stress or ultimate strength design.

CONDUIT CROSS-SECTION

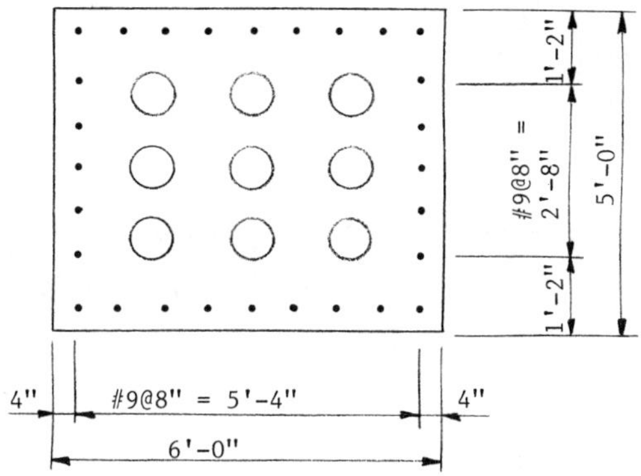

Conduit Data	Soil Data
f'_c = 3000 psi (209 kg/cm^2)	Uniform depth of soil Depth = 30 ft (9.2 m) Conduit depth (℄) = 10 ft (3.1 m)
*E_c = 5×10^6 psi (3.5×10^5 kg/cm^2)	
E_s = 29×10^6 psi (2.0×10^6 kg/cm^2)	Shear wave velocity = 700 fps (213 m/sec)
f_y = 60000 psi (4186 kg/cm^2)	Poisson's ratio = 0.30
A = 30.0 ft^2 (2.7 m^2)	
I_H = 90.0 ft^4 (0.78 m^4)	Shear Modulus = 1.9×10^6 psf (920500 kg/cm^2)
I_V = 62.5 ft^4 (0.54 m^4)	
	Cohesion = 400 psf (0.2 kg/cm^2) Friction angle = ϕ = 15° Sat. weight = γ_{sat} = 125 pcf (2 g/cm^2) Moist weight = γ = 120 pcf (1.9 g/cm^2) Water Table = 15 ft (4.6 m) down

*E_c is not based on ACI 318-77 where $E_c = W_c^{1.5}\ 33\sqrt{f'_c}$, but is based on reasonable value expected in the long-term. The values used for E_c should be verified by tests.

The basis for the rebar layout is temperature and shrinkage and other loads. Here again the rebar could be based on the function of the structure and decided based on judgment as the minimum rebar required to hold the concrete together.

Axial loads:

$$\varepsilon_a = \pm \frac{V}{C} = \frac{12}{2500 \times 12} = 4 \times 10^{-4} \text{ in/in}$$

Assume concrete cracks and only the rebar carries axial load

$$F_a = 0.0004\ (28)(29 \times 10^6) = 325 \text{ kips } (146250 \text{ kg})$$

This is less than friction force times one quarter of the wave length, therefore use

$$F_a = 325 \text{ kips } (146250 \text{ kg})$$

Moment and Shear Loads:

$$\varepsilon_b = \pm \frac{R_a}{(c)^2}$$

Assume resultant seismic wave is causing only horizontal motions for moment and shear.

$$\varepsilon_b = \pm \frac{36(96.6)}{(2500 \times 12)^2} = 3.9 \times 10^{-6} \text{ in/in}$$

Strain Compatible Determination of M and V:

The conduit is assumed to deflect as a sinusoidal wave. From the wave equation representing the deformed shape

$$y = \Delta' \sin\left(\frac{\pi X}{L/2}\right)$$

$$y' = \Delta' \left(\frac{2\pi}{L}\right) \cos\left(\frac{\pi X}{L/2}\right)$$

$$y'' = -\Delta' \left(\frac{2\pi}{L}\right)^2 \sin\left(\frac{\pi X}{L/2}\right)$$

$$y''' = -\Delta' \left(\frac{2\pi}{L}\right)^3 \cos\left(\frac{\pi X}{L/2}\right)$$

$$y'''' = \Delta' \left(\frac{2\pi}{L}\right)^4 \sin\left(\frac{\pi X}{L/2}\right)$$

where

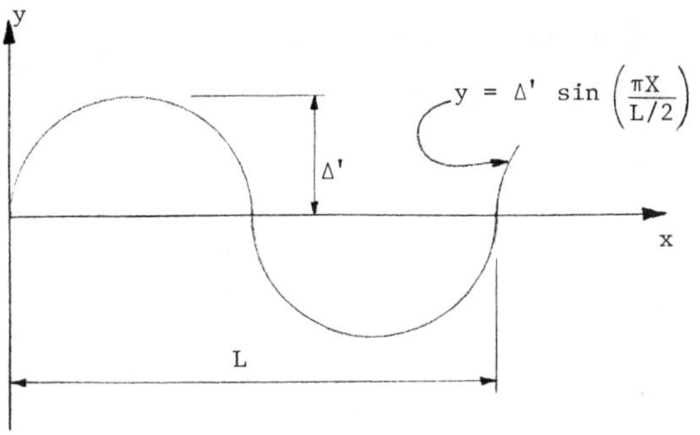

Now, moment is given by $M = EI\ y''$ and $V = EI\ y'''$, and the extreme fiber stress by $\sigma = M_{c_1}/I$. Then

$$E\varepsilon = EI\ y''\ c_1/I$$

therefore $\varepsilon = y''\ c_1$

Thus the strain in any fiber is related to the curvature (y'') and the distance from the neutral axis (c_1). Assuming the centerline of the slab moves as the wave equation dictates, then the flexural strains are symmetric about the ℄. Hence all strains are known. This assumes that the concrete does not crack. Assuming the neutral axis of the slab moves as the wave equation dictates, the curvature is known and the strain gradient is known. However, the flexural strains will change as the concrete cracks.

The governing equations for the conduit are:

Deflection $\quad y = \Delta' \sin\left(\frac{\pi X}{L/2}\right)$

Curvature $\quad y'' = -\Delta'\left(\frac{2\pi}{L}\right)^2 \sin\left(\frac{\pi X}{L/2}\right)$

Wave length $\quad L = 500$ ft (152.5 M) (Appendix 3)

$$\Delta' = \frac{M}{EI\left(\frac{2\pi}{L}\right)^2} = 0.1 \text{ in } (0.25 \text{ cm})$$

then $y''_{max} = \Delta'\left(\frac{2\pi}{L}\right)^2 = 0.008\left(\frac{2\pi}{500}\right)^2 = 1.3 \times 10^{-6}$ 1/ft
$(0.43 \times 10^{-7}$ 1/cm)

Assuming the concrete does not crack, the N.A. is at the ℄ of the conduit, then

$\varepsilon = y''c_1$ and at maximum $\varepsilon = y''_{max}c_1$

$\varepsilon_{max} = 1.3 \times 10^{-6}\ (3) = 0.000004$ in/in

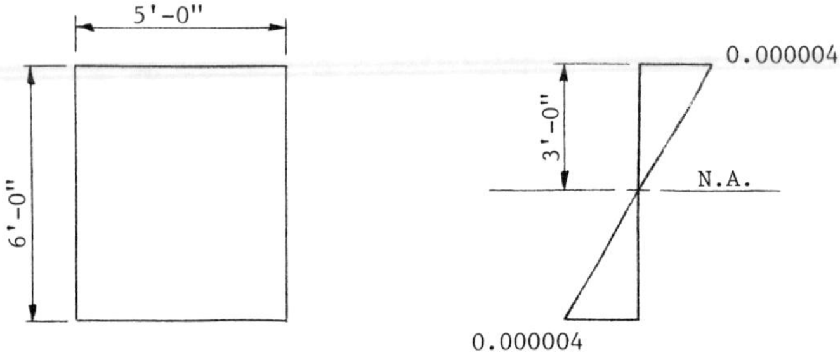

The strain is well within the allowable compressive strain for reinforced concrete.

Now assume the concrete cracks and determine the location of the N.A. To reduce the calculations, use only the rebar in the 5 ft. (1.52 m) sides of the conduit. Neglect the steel in the 6 ft. (1.83 m) sides.

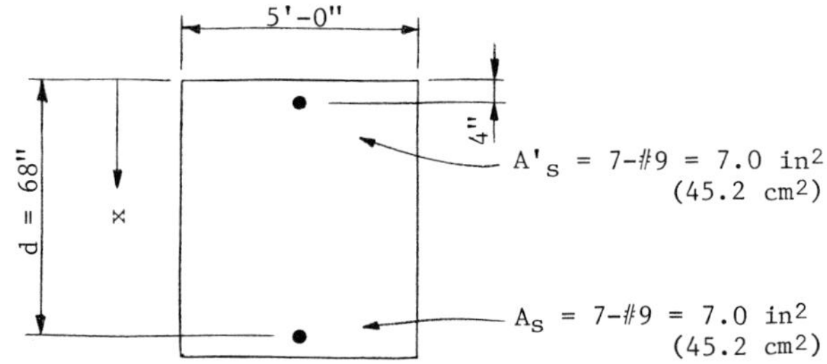

Now locate the N.A.

$$bx\left(\frac{x}{2}\right) + nA'_s\ (x - 4) = (d - x)nA_s$$

$n = E_s/E_c = 5.8$

$x = 8.47$ in. (21.5 cm)

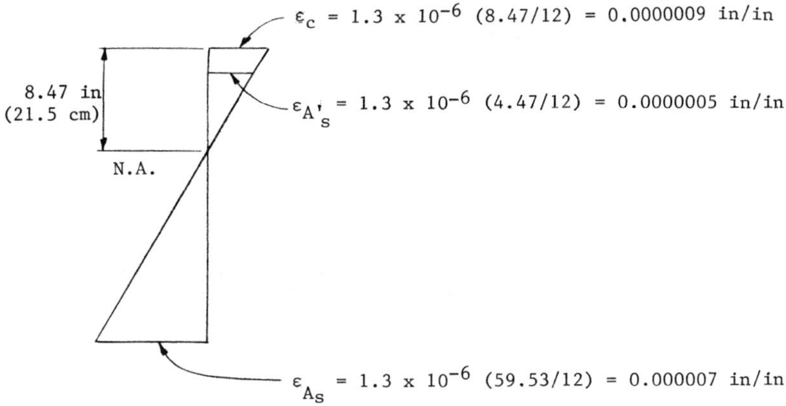

The extreme tensile strain is 0.000007 in/in. The yield strain for Grade 60 rebar is

$$\varepsilon_y = F_y/E_s = 0.00207 \text{ in/in}$$

Therefore the tension reinforcement has not yet yielded. The maximum steel stress is

$$\sigma = \varepsilon E_s = 0.000007(29 \times 10^6) = 188 \text{ psi } (13.1 \text{ kg/cm}^2)$$

Adding the stress due to the axial load gives

$$\sigma = 188 + 325000/28 = 11800 \text{ psi } (823 \text{ kg/cm}^2)$$

which is well within allowable stresses.

The extreme compression strain is 0.0000009. By working stress design the stress and strain for concrete are approximately linear up to a stress of 0.45 f'_c. For f'_c = 3000 psi this stress is

$$f_c = 0.45(3000) = 1350 \text{ psi } (94.2 \text{ kg/cm}^2)$$

and the corresponding strain is

$$\varepsilon_c = 1350/5000000 = 0.00027 \text{ in/in}$$

Therefore the compressive concrete stresses are within the linear elastic zone as assumed considering the strain due to axial load.

$$\varepsilon = 0.0000009 + \frac{325000}{30(144)(5 \times 10^6)} = 0.000016 < 0.00027$$

Thus this section can withstand the earthquake loads.

Now, based on the compression and tension forces in the section, the internal moment is

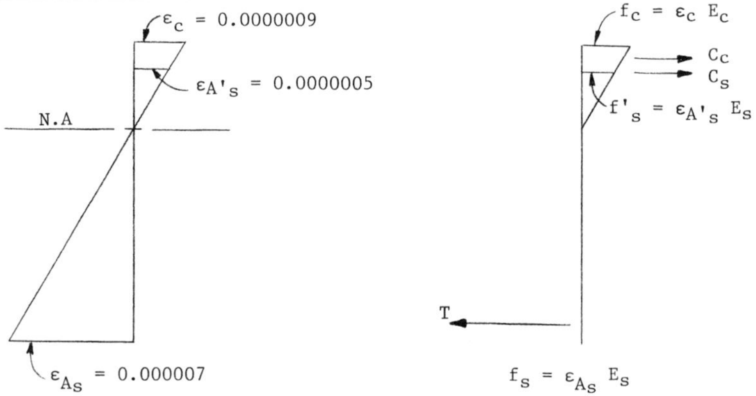

$C_c = 1/2\ f_c\ (8.47)(60) \qquad = 1145$ lbs (515 kg)

$C_s = f'_s\ A'_s \qquad = 100$ lbs (45 kg)

$T\ = f_s\ A_s \qquad = 1320$ lbs (594 kg)

$M = C\ell = T\ell = C_s(64) + C_c[68 - 1/3(8.47)]$

$= 80{,}900$ in-lb $= 6.7$ k-ft. (92470 cm-kg)

Now, calculate the cracked moment of inertia for the section

$I_{CR} = I_{CONC} + I_{A'_s} + I_{As}$

$I_{CONC} = 1/3(60)(8.47)^3 = 12{,}150\ in^4 = 0.6\ ft^4\ (0.005\ m^4)$

$I_{A'_s} = n\ A'_s d'^{\,2} = 5.8(7.0)(4.47)^2 = 810\ in^4 = 0.04\ ft^4\ (0.0003\ m^4)$

$I_{A_s} = n\ A_s\ d^2 = 5.8(7.0)(59.53)^2 = 143{,}900\ in^4 = 6.94\ ft^4\ (0.056\ m^4)$

$I_{CR} = 7.57$ ft^4 (0.061 m^4)

Based on the wave equations and the cracked moment of inertia, the moment is

$$M = EIy''_{max} = 720000(7.57)(1.3 \times 10^{-6})$$

$$= 7.1 \text{ k-ft. } (97384 \text{ cm-kg})$$

This compares well with the moment above of 6.7 k-ft (92470 cm-kg). Therefore use the cracked moment of inertia to calculate the shear, V

$$V = EI\ y''' = 720000(7.57)(1.6 \times 10^{-8}) = 0.1 \text{ k } (45 \text{ kg})$$

Thus, the following loads

$M = 7.1$ k-ft (97384 cm-kg)

$V = 0.1$ k (45 kg)

$F_a = 325$ k (146250 kg)

can be used in design with any other loads to verify the capacity of the conduit bank. If when combined with other loads, the conduit section changes, then new earthquake loads should be calculated. The design becomes an iterative process.

Appendix 5

List of Seismic Analysis Committee Members

H. J. Dahlke, Rockwell Hanford Operations
I. K. Ghosh, Brown and Root, Incorporated
E. C. Goodling, Gilbert Associates, Incorporated
R. C. Guengler, EG&G Idaho, Incorporated
A. K. Gupta, North Carolina State University
A. H. Hadjian, Bechtel Power Corporation
A. Haldar, Georgia Institute of Technology
J. F. Halsey, Brown and Root, Incorporated
W. W. Hays, US Geological Survey
R. J. Hunt, Tennessee Valley Authority
I. Idriss, Woodward-Clyde Consultants
J. J. Johnson, Structural Mechanics Associates, Incorporated
D. D. Kana, Southwest Research Institute
L. Gerdes, Combustion Engineering, Incorporated
J. K. Khanna, Flour Power Services, Incorporated
R. P. Kassawara, EDS Nuclear
R. P. Kennedy, Structural Mechanics Associates, Incorporated
L. Malik, URS/John A. Blume and Associates
D. P. Moore, Southern Company Services
B. Murray, Lawrence Livermore Laboratory
M. O'Rourke, Rensselaer Polytechnic Institute
D. V. Reddy, Memorial University of Newfoundland
*L. C. Shao, Nuclear Regulatory Commission
S. Singh, Sargent and Lundy Engineers
P. D. Smith, Lawrence Livermore Laboratory
C. V. Subramanian, General Electric Company
T. Blejwas, Sandia National Laboratories
R. Wheaton, EDS Nuclear
P. Yanew, EQE, Incorporated

*Corresponding Member